FROM CRITICAL SCIENCE TO SOLUTIONS

The Best of Scientific Solutions

Edited by
Richard Clapp

Boston University School of Public Health
and
University of Massachusetts-Lowell

Work, Health, and Environment Series
***Series Editors*: Charles Levenstein, Robert Forrant, and John Wooding**

Routledge
Taylor & Francis Group
LONDON AND NEW YORK

First published 2012 by Baywood Publishing Company, Inc.

2 Park Square, Milton Park, Abingdon, Oxfordshire OX14 4RN
605 Third Avenue, New York, NY 10017

Routledge is an imprint of the Taylor & Francis Group, an informa business

First issued in hardback 2020

Library of Congress Catalog Number: 2011017705

ISBN 13: 978-0-89503-404-5 (hbk)
ISBN 13: 978-0-89503-408-3 (pbk)

Library of Congress Cataloging-in-Publication Data

From critical science to solutions : the best of scientific solutions / edited by Richard Clapp
p. cm. -- (Work, health, and environment series)
Includes bibliographical references and index.
ISBN 978-0-89503-404-5 (cloth : alk. paper) -- ISBN 978-0-89503-408-3 (pbk. : alk. paper) -- ISBN 978-0-89503-467-0 (epub) -- ISBN 978-0-89503-468-7 (epdf) 1. Environmental health. 2. Science. I. Clapp, Richard, 1945-
RA565.F76 2011
362.1--dc23

2011017705

Cover Art by Christopher R. Lindstrom

Table of Contents

Section III
SOLUTIONS SCIENCE

Preface

In the Baywood series *Work, Health, and Environment,* the conjunction of topics is deliberate and critical. We begin at the point of production—even in the volumes that address environmental issues—because that is where things get made, workers labor, and raw materials are fashioned into products. It is also where things get stored or moved, analyzed or processed, computerized or tracked. In addition, it is where the folks who do the work are exposed to a growing litany of harmful things or are placed in harm's way. The focus on the point of production provides a framework for understanding the contradictions of the modern political economy.

Despite claims to a post-industrial society, work remains essential to all our lives. While work brings income and meaning, it also brings danger and threats to health. The point of production, where goods and services are produced, is also the source of environmental contamination and pollution. Thus, work, health, and environment are intimately linked.

Work organizations, systems of management, indeed the idea of the "market" itself, have a profound impact on the handling of hazardous materials and processes. The existence or absence of decent and safe work is a key determinant of the health of the individual and the community: what we make goes into the world, sometimes improving it, but too often threatening the environment and the lives of people across the globe.

We begin this series to bring together some of the best thinking and research from academics, activists, and professionals, all of whom understand the intersection between work and health and environmental degradation, and all of whom think something should be done to improve the situation.

The works in this series stress the political and social struggles surrounding the fight for safer work and protection of the environment, and the local and global struggle for a sustainable world. The books document the horrors of cotton dust, the appalling and dangerous conditions in the oil industry, the unsafe ways in which toys and sneakers are produced, the struggles to link unions and communities to fight corporate pollution, and the dangers posed by the petrochemical industry, both here and abroad. The books speak directly about the contradictory effects of the point of production for the health of workers, community, and the environment. In all these works, the authors keep the politics front and foremost. What has emerged, as this series has grown, is a body of scholarship uniquely focused and highly integrated around themes and problems absolutely critical to our own and our children's future.

About This Volume

In this volume, I collected articles that primarily appeared in the "Scientific Solutions" section of *New Solutions, A Journal of Environmental and Occupational Health Policy*. This section began in the very first issue in Spring 1990 with an introduction by Dr. Michael Silverstein, who first edited the section. Following the inspiration of Tony Mazzocchi, who founded the journal and got the Oil, Chemical and Atomic Workers International Union to publish it, Mike Silverstein called for articles that were unembarrassed about urging strong action to protect workers' health even in the absence of definitive cause-effect scientific knowledge. Some of the first Scientific Solutions articles responded admirably to this call, as did some solicited by the second section editor, Dr. Ellen Silbergeld, and have been collected elsewhere.

I became editor of Scientific Solutions in 1994, and have selected articles that were published between 1994 and 2010. I have grouped the articles into three general categories, starting with what I have called Critical Science. These articles are primarily critiques of "how science is done" or how science is incorporated into public health policy in the United States and elsewhere. Some of these, such as the critique of radiation epidemiology by Wing, Richardson, and Stewart, take issue with the fundamental assumptions of normal science in the field. The second category is what I have called Precautionary Science. These articles are essentially what Mike Silverstein originally called for in his Introduction to Scientific Solutions. These articles, such as the ones by Dement on asbestos and Solomon, and colleagues on the risks of manganese, essentially call for precautionary regulations to reduce exposures where there is substantial but, in the eyes of some, less than definitive scientific knowledge.

The final category is what I have labeled Solutions Science. In some ways, this represents the current stage of precautionary science, where we have begun to look at larger societal issues and have moved beyond traditional scientific approaches and critiques. Barry Commoner's article, which he presented at a festschrift honoring his contributions, asserts that the problems of environmental degradation and related health effects will not be solved until the fundamental problem of poverty and inequality on a global scale has been addressed. Another article in the Solutions Science section deals with advancing a new approach to chemicals policy in California, following the principles of Green Chemistry that have been popularized by Anastas and Warner in the past decade or more. This collection ends with an essay that has not been published elsewhere, but which was developed by colleagues at the University of Massachusetts Lowell, in collaboration with the first author. This chapter describes the current thinking of a group of several contributors to *New Solutions,* although it does not address the political and economic constraints on progressive science in the coming period. That will be the task of the next generation of advocates for worker health and environmental health.

INTRODUCTION

When Science Doesn't Have All the Answers

Editor, Michael Silverstein*

The interface between science and public policy is often occupied by tortured discussions about the certainty of results and conclusions. The scientific literature reflects just how uncomfortable many people have become with this. Articles tend to be of two types. In the first the author is able to convince him or herself that the science is conclusive and that therefore specific public health recommendations are warranted. In the second, more common, situation the author pleads uncertainty and urges no more than further study.

Those who fall prey to the belief that there are only these two mutually exclusive options are liable to end up taking positions which undermine their own credibility. Progressive public health activists have sometimes overstated the strength and consistency of positive findings in the mistaken belief that scientific certainty can be the only justification for advocating active protective policy measures. Those more responsive to traditional academic pressures, on the other hand, may use the convenient shield of statistical significance to avoid the stigma of having made recommendations on the basis of results which may later prove to have been falsely positive. An extreme manifestation of the policy paralysis adopted by many scientists confronting these questions has been embodied in the editorial practice of another new journal which has declared that it will not publish scientific articles which include policy recommendations [1].

Our intent in the "Scientific Solutions" section of *New Solutions* will be to publish articles which are unembarrassed about urging strong action on the basis

*Editor, Michael Silverstein wrote this Introduction in *New Solutions,* 1990.

of less-than-definitive science, but which also explicitly acknowledge gaps, uncertainties and disputes about the knowledge base. Our rationale is simply that scientific certainty and unanimity are so rare that public health decision-making must proceed when science is incomplete, uncertain, or controversial. As Richard Monson has written, "Whereas opinions as to the meaning of any association seen in epidemiologic data will be varied, action to be taken based on the data must be definitive. Behavior will either be altered or it will stay the same. There is no room for equivocation" [2]. Similarly, the U.S. Supreme Court stated in its benzene decision that the Occupational Safety and Health Administration (OSHA) would be straightjacketed if it were required to demonstrate scientific certainty before undertaking regulatory action [3]. The intent of the "Scientific Solutions" section will be to generalize the logic of these statements and to apply it to a variety of public health topics.

REFERENCES

1. K. Rothman, *Epidemiology* Journal, Guidelines for Contributors.
2. R. Monson, *Occupational Epidemiology,* CRC Press, Boca Raton, Florida, p. 93, 1980.
3. *Industrial Union Department, AFL-CIO v. American Petroleum Institute* 448 U.S. 607, 100 S. CT. 2844 (1980).

Section I

CRITICAL SCIENCE

CHAPTER 1

A Case Study of Pseudo-Science in Occupational Medicine

Sven Ove Hansson

A detailed analysis is performed of a paper in which H. E. Stokinger, for many years the president of the American Conference of Governmental Industrial Hygienist's Threshold Limit Values Committee, purports to show that eight specified chemical carcinogens have thresholds. For seven of the eight substances, his sources do not at all support his conclusions, and in some of these cases they even support the opposite conclusion. Even in the remaining case, bis(chloromethyl)ether, the results that he refers to are compatible not only with a threshold but also with various non-linear non-threshold models. Stokinger's paper satisfies classical criteria of pseudo-science, such as the misrepresentation of data, incorrect statistics, reliance on anecdotal evidence, the use of scientific references in support of claims that they do not at all substantiate, and the uncritical acceptance of information coming from sources known to be fraudulent.

In 1987, Herbert E. Stokinger published a paper entitled "Nonstatistical vs. illusory statistical approaches to the estimation of risk from environmental chemicals." The paper purports to "dispel a few unbecoming statistical notions." According to its author, we should "[a]bandon the practice of statistical tampering with the data and the endless fruitless search for models," since "[e]ndpoints of effect and no-effect levels are clearly evident on inspection" so that not very much statistics is needed [1, pp. 2-3]. Most of the paper is devoted to an attempt to show that chemical carcinogens have thresholds, that is, non-zero dose levels at which they do not give rise to cancer. Stokinger quotes evidence relating to eight carcinogens, evidence which he claims to provide strong support for the existence of thresholds in chemical carcinogenesis (henceforth: the threshold hypothesis). A typical example of his argumentation is that since vinyl chloride has been shown to cause cancer at 50 ppm, but not at 10 ppm, there is evidence of "a threshold somewhere below 50 and above 10 ppm" [1, p. 6].

Stokinger was a member of the Threshold Limit Values (TLVs) committee of the American Conference of Governmental Industrial Hygienists (ACGIH) for twenty-six years and its chairman for fifteen years [2, p. 10]. Due to the immense worldwide impact of the TLVs set by this body, he has been one of the more influential persons in modern regulatory toxicology [3]. This is reason enough for a close examination of his views on chemical carcinogenesis.

After the methodology of the present study has been introduced in Section 1, a detailed study of Stokinger's sources follows in Section 2. In Section 3, the general nature of his evidence is summarized, and the perhaps somewhat provocative title of the present chapter is accounted for.

1. METHODOLOGY

A close reading of Stokinger's sources was undertaken in order to answer three questions:

1. *What types of publications does he refer to?* In particular, have they passed a peer-review process? One should expect a scientifically well-founded argument in favor of the threshold hypothesis to be based primarily on final reports of experimental and/or epidemiological studies that have been published or accepted for publication in peer-reviewed journals.
2. *What are the institutional affiliations of the scientists whose work he refers to?* The ACGIH has been criticized for relying too much on the laboratories and medical officers of companies that produce the substances under review [4]. It is therefore of interest to see whether there is any bias in this direction among the references on which Stokinger bases his argument.
3. *To what degree do Stokinger's sources support his conclusions?* In answering this question, I will focus on the statistical issues, thus giving Stokinger the benefit of the doubt with respect to biological relevance. In particular, I will not question the relevance to human health of the results from rodents to which he refers.

As mentioned above, Stokinger's arguments typically refer to two exposure levels such that chemically induced cancer appeared at the higher but not at the lower level. From this, he concludes that there is a threshold somewhere between these two levels.

This type of argument is problematic from a statistical point of view, as can be seen from the following hypothetical example:

Example 1: One hundred rodents are exposed to 50 ppm of a substance, and six of them develop an unusual form of cancer such as angiosarcoma.

Another 100 rodents of the same species, strain, and sex are exposed to 5 ppm, and none of them contracts the disease.

Such an experiment would not provide any evidence in favor of the threshold hypothesis. To see this, let us make the common assumption that if there is no threshold, then at least for low doses the incidence of cancer may be approximately proportionate to the level of exposure. In other words, let us test the following *linear hypothesis:* If the (observed) incidence of (some type of) chemically induced cancer is f at the lowest dose level D for which carcinogenesis has been demonstrated, then the (true) frequency of chemically induced cancer at any dose level $D' < D$ is $f \times (D'/D'/D)$. This linear hypothesis is (a simplified version of) the major rival hypothesis against which Stokinger is arguing. Therefore, only data that speak against the linear hypothesis can be taken as evidence in favor of the threshold hypothesis. Data that are compatible with both hypotheses cannot be used in favor of either of them.

Testing the Threshold Hypothesis

In order to test the threshold hypothesis in our example, we should assume that the true incidence of exposure-related cancer at 5 ppm is 5/50 = 1/10 of the observed incidence at 50 ppm (0.06), that is, that it is 0.006. With that incidence, to find 0 cases among 100 exposed is no surprise. To the contrary, the probability is 55 percent that this will happen. The observed outcome is expected according to the linear hypothesis, and thus cannot be taken as an argument in favor of the threshold hypothesis.

A result with no chemically induced cancer at the lower of the two dose levels will be labeled as *expected according to the linear hypothesis* if the probability of this outcome is at least 50 percent under the assumption that the linear hypothesis holds.

Next, consider the following modified version of the example:

> *Example 2:* One hundred rodents are exposed to 50 ppm of a substance, and twenty of them develop angiosarcomas. Another 100 animals are exposed to 5 ppm, and none of them contracts the disease.

According to the linear hypothesis, the expected incidence of angiosarcoma among animals exposed to 5 ppm is 0.02. The probability, given this incidence, of no angiosarcomas among the exposed 100 animals is 0.13. According to the common criterion of statistical significance in biomedical research ($p < 0.05$), this outcome does not represent a significant deviation from the linear hypothesis.

A result with no chemically induced cancer at the lower of the two dose levels will be labeled as *compatible ($p \geq 0.05$) with the linear hypothesis* if the probability of this outcome is at least 5 percent under the assumption that the linear hypothesis holds. Finally, such a result will be said to *deviate significantly* *from the linear hypothesis* if its probability is less than 5 percent under the assumption that the linear hypothesis holds.

2. FINDINGS

Stokinger's evidence refers to eight substances: bis(chloromethyl)ether, 1,4-dioxane, coal tar, β-naphthylamine, HMPA, vinyl chloride, dimethyl sulfate, and asbestos.

Bis(chloromethyl)ether (BCME): Stokinger's reference is an abstract published in 1975 by B. K. J. Leong and co-workers at the laboratory of Dow Chemical Company [5]. According to the abstract, ninety-eight out of 111 rats exposed for six months to 100 ppb of the substance had developed esthesioneuroepitheliomas, whereas no such tumors had been noticed in groups exposed to 10 or 1 ppb. (The experiment was still ongoing.) The number of animals in the latter groups was not mentioned.

The final report of the experiment was published in 1981 [6]. (In his 1987 paper, Stokinger only referred to the abstract from 1975.) Still, no tumorigenic response had been observed at 10 or at 1 ppm. The authors claimed that their data "demonstrated the existence of a non-tumorigenic or no-observable-effect-level of BCME vapor for a six-month inhalation exposure in rats and mice" [6, p. 269]. The number of animals examined after exposure to 10 ppb was now reported to be 111. Under the linear hypothesis, the expected frequency at this exposure level is 0.088. The obtained zero result deviates significantly from the linear hypothesis ($p = 0.00004$).

1,4-Dioxane: For this substance, Stokinger referred to two results from the laboratories of the Dow Chemical Company.

- 0.1% of the substance in drinking water gave rise to no tumors in rats.

As can be seen from the report that Stokinger referred to here [7], tumors were observed at 1 percent but not at 0.1 percent. Each exposure group consisted of 120 animals. Thirteen tumors in the 1 percent group were regarded as treatment-related (10 hepatocellular carcinomas and 3 nasal carcinomas). Hence, according to the linear hypothesis, the expected incidence of such tumors in the 0.1 percent exposure group is 13/1,200. To observe no such tumor in this group is compatible with the linear hypothesis ($p = 0.27$), and thus does not support the threshold hypothesis. (In fact, one hepatocellular carcinoma each was observed in the 0.1 percent group and the 0 percent control group, but this does not strengthen the case for the threshold hypothesis.)

- In rats exposed by inhalation to 111 ppm of the substance, "no hepatic or nasal carcinomas were observed, although they have been reported to occur in rats maintained on water containing high levels of dioxane" [8, p. 287]. Stokinger sees this as evidence of "a threshold somewhere between this level and below 1,000 ppm" [1, p. 6].

No Conclusion Can Be Drawn

This result was based on pathological examination of 423 exposed animals. Two hundred and eighty-nine controls were examined, also with no hepatic or nasal carcinoma found. However, no exposures other than 0 and 111 ppm were reported. Due to the absence of information about tumor incidence at higher exposure levels, no conclusion with respect to the threshold hypothesis can be drawn from this report. No explanation of Stokinger's upper limit 1,000 ppm could be found either in the papers that he refers to or in other studies on the substance reported in the open literature.

Coal Tar: For this substance, Stokinger referred to a National Institute for Occupational Safety and Health report from 1973 by E. Bingham [9]. According to his summary, "application of coal tar pitch volatiles to mouse skin resulted in tumors at total doses of 6,400, 640, and 64 mg, but not at three doses below 64 mg" [1 = p. 6]. In his table, he gave "<0.64 mg" in the column for "dose levels not eliciting tumors" [1, p. 5].

Bingham's report does not contain exactly the information that Stokinger ascribed to it. In particular, the dose levels mentioned by Stokinger do not appear in the report. However, there were three exposure groups without tumors. It can be assumed that Stokinger referred to these three groups.

- None of sixty-six mice exposed to a total dose of 0.048 mg B(a)P in toluene developed tumors.

In comparison, seven out of eighty-three mice exposed to a total dose of 0.48 mg developed tumors. According to the linear hypothesis, we should expect the tumor frequency at 0.048 mg to be 7/830. Under this assumption, the probability of finding no tumors among sixty-six animals is 57 percent, so that the result obtained is expected according to the linear hypothesis.

- None of sixty mice exposed to 0.0048 mg B(a)P in toluene developed tumors.

According to the linear hypothesis, and applying the information from the 0.48 mg B(a)P level already mentioned, we should expect the incidence of tumors in the 0.0048 group to be 7/8,300, and the probability of finding no tumors among sixty animals is 95 percent (sic). Hence, this result is (highly) expected according to the linear hypothesis.

- None of fifty-six mice exposed to 0.8 mg coal tar in toluene developed tumors. In contrast, thirty-nine out of seventy-six animals exposed to 8 mg of the same substance developed tumors.

Result (Barely) Compatible

According to the linear hypothesis, the expected tumor incidence at 0.8 mg is 39/760. The probability of finding no tumor among fifty-six animals is 5.2 percent, so that the obtained result is (barely) compatible with the linear hypothesis.

For β-naphthylamine, Stokinger's only reference is "Private communication from J. A. Zapp." (Zapp was at the time an employee of the du Pont company [10].) The contents of this information is: "From the information available to 1979, thirty-six years of exposure have elapsed without the appearance of bladder tumors in workers exposed to β-naphthylamine containing less than 0.5 percent of the β derivative, whereas tumors occurred in prior exposures to β-naphthylamine containing more than 5 percent β derivative" [1, p. 6]. In Stokinger's list of references, this information is said to have been communicated in 1975 (sic) [1, p. 8].

I have not been able to locate any publication by Zapp in which the relevant details, such as the number of exposed workers and the levels and durations of exposures, are given. In the absence of such information, nothing can be concluded about possible thresholds. (Note that there was more than a 10-fold reduction in the concentration of β-naphthylamine in the product. Additional reductions in exposure may have taken place due to improved technology. If the total reduction in exposure to β-naphthylamine was large, the linear model may account for a large reduction in tumor incidence.)

For *hexamethyl phosphoramide* (HMPA), Stokinger referred to a preliminary report from 1975 by J. A. Zapp [10]. (Stokinger erroneously cites the paper as covering seven pages in the journal, 916-922. It is contained in the single page 916.) According to Stokinger, Zapp's paper includes evidence from both rodents and humans, supporting the existence of a threshold for this carcinogen.

- Rodents: HMPA, Stokinger said, "induces squamous cell carcinomas in the nasal cavity of rats exposed daily for three months by inhalation at 0.05 ppm, and after a 13-month period. Daily exposures at 0.01 ppm for 24 months, however, elicited no tumors . . ." [1, p. 6].

In his paper, Zapp reported results from the first eight months of a study including "four groups of 120 male and 120 female rats each, exposed to 0, 50, 400, and 4,000 ppb of HMPA respectively, exposed for six hours a day, five days a week. A planned six-month sacrifice of eighteen animals from each group showed no compound-related lesions." In the eighth month, several animals in the two highest exposure groups had developed squamous cell carcinoma in the nasal cavity. "The incidence was dose-related, twelve nasal tumors being found in fifteen rats dead or sacrificed in extremis, out of 240 exposed, in the eighth month in the high exposure group (4,000 ppb) as compared with seven tumors in eight dead or sacrificed in extremis, out of 240 exposed, in the mid-exposure group

(400 ppb). Six rats were randomly selected from the low-dose (50 ppb) for sacrifice and no malignancy was found" [10].

Details at Variance with Paper

As we can see, the details concerning levels and durations of exposure given by Stokinger are at variance with the paper to which he refers. Applying the linear hypothesis to Zapp's report, we should assume the incidence of tumors at 50 ppm to be one-eighth of the incidence at 400 ppb, that is, it should be 7/1,920. Under this assumption, the probability of finding no tumor among 240 animals exposed to 50 ppb is 42 percent. Hence such a report is compatible with the linear hypothesis (and close to being expected according to it).

Since 1982, final reports have been available from this experiment [11-13]. In these reports, a group exposed to 10 ppm is also included. Among animals exposed to 50 ppm, 20 percent developed tumors, whereas none of 200 animals exposed to 10 ppm developed tumors (after 24 months). This result deviates significantly from the linear hypothesis. However, this information is not to be found in the paper to which Stokinger refers.

- Humans: Stokinger said that ". . . nor have neoplasias appeared in workers after more than twenty years of Kelvar fiber production, indicating a probable threshold for man as well as for animals" [1, p. 6].

Not a single word about human exposure can be found in the paper by Zapp which Stokinger refers to in support of this statement. (Neither can it be found in the other brief communication on the subject that Zapp published in the same year [14].)

Vinyl Chloride: Stokinger claimed that there are three types of evidence for a threshold somewhere between 10 and 50 ppm: 1) human evidence, 2) animal evidence referring to cancer incidence, and 3) animal evidence of a biochemical nature.

1. The human evidence is that "more than 35 years have passed since workers were exposed to vinyl chloride containing small amounts (5 ppm) of vinylidene chloride, without the appearance of tumors. Vinyl chloride levels during the 1950s, although averaging 160 ppm, rose occasionally to 1,000 ppm at some work operations, but later were around 50 ppm or below. This gives epidemiological evidence of a threshold somewhere below 50 and above 10 ppm" [1, p. 6]. The reference given for this statement is a paper from 1972 by C. G. Kramer and T. E. Mutchler of the Dow Chemical Company [15].

Kramer's and Mutchler's findings refer to a group of ninety-eight male workers who had been exposed to vinyl chloride "for periods up to 25 years" [15, p. 19] (not the "more than 35 years" mentioned by Stokinger). No tumors had developed in the group. However, no mention is made of the tumor incidence at any higher dose level. Without such information, no indication of the plausibility of a threshold can be obtained.

2. Stokinger referred to two animal studies "in which a few tumors were still appearing in rats at 50 ppm, but none at 10 ppm, indicating a threshold somewhere between 50 and 10 ppm" [1, p. 6]. One of these studies is a paper by M. L. Keplinger from 1975 [16] and the other a paper referred to as "Maltoni." No paper by Maltoni appears on the list of references, but in another paper [17] Stokinger referred to a paper by C. Maltoni and G. Lefemine from 1975 [18] in a similar context.

Faked Toxicity Data

The paper by Keplinger and co-workers reports tests carried out at the Industrial Bio-Test Laboratories. In 1983, three officials of this laboratory were convicted of mail fraud. Their crime was to have faked toxicity data on drugs under review by the Food and Drug Administration and the Environmental Protection Agency [19]. One of the three convicted officials was Moreno Keplinger.

Stokinger, in his 1987 paper, does not mention the problematic status of experiments reported by Keplinger.

At any rate, the paper by Keplinger et al. does not contain the information ascribed to it by Stokinger. It describes a study in which rats, mice, and hamsters were exposed to 0, 50, 200, and 2,500 ppm vinyl chloride. No animals were exposed to 10 ppm. The paper is a brief preliminary report. (Incidentally, it covers less than one and a half pages, not the six pages, 219-224, assigned to it in Stokinger's list of references.)

Angiosarcomas had been discovered at 50 ppm, as Stokinger says, but of course no information about effects at 10 ppm can be gained from this study.

The report by Maltoni and Lefemine summarizes a long series of experiments conducted by the authors at the Istituto di Oncologia and the Centro Tumori in Bologna, Italy. A carcinogenic effect had been shown at 50 ppm. No results from exposures lower than 50 ppm were reported. However, a study with exposures to 25, 10, and 5 ppm was described as "just started" [18, p. 199]. In 1981, a final report from this study was published. Among the findings were five liver angiosarcomas out of 120 animals at 25 ppm, and one liver angiosarcoma out of 120 animals at 10 ppm [20, p. 20]. Hence, Stokinger was wrong when he claimed in 1987 that this experiment showed that no tumors appeared at 10 ppm.

3. Finally, Stokinger quoted a paper from 1976 by P. G. Watanabe and co-workers at the Dow Chemical Company [21]. According to Stokinger, this paper provides "indisputable evidence of a threshold for vinyl chloride, although in rats." More precisely, the paper is said to indicate "a threshold for vinyl chloride hemangiomas (tumors) of the rat liver somewhere more than 10 ppm and less than 50 ppm." In this experiment, hepatic non-protein sulfhydryl was used as an indicator of the body's capacity to neutralize vinyl free radicals. "[A] distinct dose-response relation was found for exposure levels of 2,000, 1,000, 250, 150, 50, and 10 ppm. Exposure at the four highest levels caused a progressive

depression of the hepatic nonprotein sulfhydryl content, whereas no depression was observed at 10 ppm" [1, p. 6].

Stokinger's description of the experiment is correct, but it leaves out two important facts: The exposures were short, and the groups were small. The information about effects at 10 ppm was based on five animals that had been exposed for seven hours. No statistically significant effect was discovered with the Student *t* test ($p < 0.05$). Clearly, a *biologically* significant effect at 10 ppm might have gone undetected in this very small experiment.

Dimethyl Sulfate (DMS): Stokinger gave two references in support of a threshold for the carcinogenicity of this substance. Both relate to human exposure.

1. An epidemiological study by Pell from 1972 is said to have shown "no excess incidence of cancer of the respiratory tract in workers exposed up to 26 years at levels frequently well above 1 ppm" [1, p. 6]. The reference given for this is "S. Pell, Sept 1972. Epidemiological study of dimethyl sulfate and cancer of the respiratory system. E.I. Du Pont & Co., Wilmington, DE" [1, p. 8]. This study does not seem to have been reported in the open literature. I have not been able to locate it.

2. "Similarly, no overt cases of cancer of the lung attributable to DMS occurred in a German plant where workers receive annual physical examinations" [1, p. 6]. For this information, Stokinger referred to a paper by H. Druckrey and co-workers (Max-Planck-Institut) from 1970 [22]. This paper, however, is concerned exclusively with experimental studies and does not provide the epidemiological evidence that Stokinger attributes to it. Incidentally, it is the third part of a series of papers. The first part reports a case study in which DMS was found to be the probable cause of bronchial carcinoma in several heavily exposed workers [23].

Got His References Wrong

In all probability, Stokinger just got his references wrong. In an earlier paper [17], he referred to the same information as obtained in a "Letter communication to Dr. John A. Zapp from Dr. H. Zeller, Med. Dir. Badische Anilin and [sic] Soda Fabrik (Apr. 13, 1972)." I have not been able to locate any published information that can substantiate this claim.

Asbestos: Stokinger referred to two papers from 1973 by P. E. Enterline and co-workers (University of Pittsburgh) [24, 25]. "This report concluded that important increments in respiratory cancer mortality apparently occur somewhere between 100 and 200 mppcf-years exposure, and that no direct relationship between asbestos dust exposure and respiratory cancer occurred below 125 mppcf years. This conclusion was based on an epidemiological study of the relationship of respiratory cancer to occupational exposures among 1,348 retired asbestos workers, who had an average length of employment of 25 years, and often at very high dust exposures" [1, p. 7]. In a table, Stokinger gives "<125 mppfc/yrs" in the column for "dose levels not eliciting tumors" [1, p. 5].

The papers by Enterline and his colleagues report a study of 1,348 retired asbestos workers. Among these were a group of 533 men with a total dust exposure below 125 mppcf-yr. The average exposure in this group was 62.9 mppcf-yr. They exhibited a 67 percent excess of deaths in respiratory cancer (15 observed, 9 expected). However, when this group was divided into subgroups, the relationship between exposure and respiratory cancer was quite disorderly. (There was a 54 percent excess in death due to respiratory cancer among those with less than 25 mppcf-yr, a 158 percent excess in the interval 25-62.4 mppcf-yr and only an 8 percent excess in the interval 62.5-124.9.) Clearly, this lack of a dose-response relationship is puzzling, and the authors cautiously conclude that it "cannot definitely be established from these data" whether the 67 percent excess in the whole group is related to asbestos dust or to something else [24, p. 165]. Stokinger's claim that this study has shown levels below 125 mppcf-yr to be "dose levels not eliciting tumors" severely misrepresents data. (See also the discussion in [26] of Enterline's work.)

3. SUMMARY AND DISCUSSION

The results from the previous section are summarized in Table 1. As can be seen from the third column, with one or two exceptions the references date from the period 1972-1976. The classification of the references in the fourth column can be summarized as follows:

Journal, full paper	5
Journal, abstract or preliminary report	5
Unpublished report	2
Private communication	2

The institutional affiliations of the researchers referred to by Stokinger are seen from the fifth column of the table. They can be summarized as follows:

Chemical industry	10
Private laboratory	1
Universities	2
Government agency	1

Hence, Stokinger's choice of references is biased in a remarkable way toward researchers affiliated with the chemical industry.

The sixth column records the degree to which Stokinger's sources substantiate his claims. In the cases marked "??," the paper referred to by Stokinger did not at all contain the information that he claims to have found there. Two papers said by him to contain information about exposure of rodents to 10 ppm vinyl chloride contain no such information at all. He also attributes information about human exposure to HMPA to a paper by Zapp and about human exposure to dimethyl sulfate to Druckrey et al., but neither paper contains any information whatsoever about human exposure.

In the cases marked "?," the papers referred to by Stokinger described studies of the types attributed to them, but these papers do not contain sufficient information to give any indication whatsoever of the validity of the threshold or the linear hypothesis.

Only in five cases did the papers referred to by Stokinger contain sufficient information to determine whether or not the results are supportive of the threshold hypothesis.

In one of these five cases, asbestos, the report referred to provides strong evidence of carcinogenicity below the level that Stokinger claims to be a threshold. Enterline's report [24], with its evidence of a 67 percent excess death rate in respiratory cancer among workers exposed to less than 125 mppcf-yr is cited by Stokinger as "evidence" that there is no excess cancer below 125 mppcf-yr.

Compatibility Misreported

In three additional cases, the results referred to by Stokinger as supportive of the threshold hypothesis were in fact compatible with the linear hypothesis. In only one of the five cases, bis(chloromethyl)ether, did the results reported by Stokinger deviate significantly from the linear hypothesis. This is the only one of his many references that can at all be said to support the threshold hypothesis.

However, even this is a very weak support of the threshold hypothesis. The reason for its weakness is that results contradicting the linear hypothesis are compatible with various non-linear dose-response models without thresholds. Thus, the findings reported in [5] for bis(chloromethyl)ether are statistically compatible with a model that sets the tumor incidence at 10 ppb to 1/35 or less of the incidence at 100 ppb.

In conclusion, Stokinger's argumentation exhibits classic signs of pseudo-science, such as the misrepresentation of data, incorrect statistics, reliance on anecdotal evidence, the use of scientific references in support of claims that they do not at all substantiate, and the uncritical acceptance of information coming from sources known to be fraudulent. It is disturbing to find indications that pseudo-scientific practices may have played a role in issues so important to human health as the setting of occupational exposure limits.

Needless to say, although Stokinger's argumentation does not give us reasons to believe that these carcinogens have thresholds at the levels he indicates, they may nevertheless have thresholds at these or other levels. However, a scientifically well-founded argument for the threshold hypothesis cannot be based primarily on the demonstration that no cases of cancer have been detected at certain exposure levels. Instead, it will have to be based on a much more sophisticated understanding of toxicological mechanisms.

Table 1. Stokinger's Purported Evidence for a Threshold for Chemical Carcinogenesis

(1) Substance, Species, Route of Exposure	(2) Stokinger's Claim	(3) Reference, Year	(4) Type of Publication	(5) Institutional Affiliation	(6) Degree of Support for Stokinger's Claim
Bis(chloromethyl)ether Rats, inhalation	Threshold between 10 and 100 ppm.	[5] 1975	journal, abstract	Dow Chemical Company	+
1,4-dioxane Rats, ingestion	Threshold between 0.1% and 1% in drinking water.	[7] 1974	journal, full paper	Dow Chemical Company	–
1,4-dioxane Rats, inhalation	Threshold between 111 and 1000 ppm.	[8] 1974	journal, full paper	Dow Chemical Company	?
Coal tar Mice, skin application	Three low dose levels are all below the threshold.	[9] 1982	unpublished report	NIOSH	- -, - -, -[4]
β-naphthylamine Humans, inhalation	Threshold between two reported industrial exposure levels.	1975?	private communication	du Pont Company	?
HMPA Rats, inhalation	Threshold between 0.01 and 0.05 ppm.	[10] 1975	journal, prel. report	du Pont Company	- -[1] (+[3])
HMPA Humans, inhalation	Actual exposure levels are below threshold.	[10] 1975	journal, pref. report	du Pont Company	??
Vinyl chloride Humans, inhalation	Threshold between 10 and 50 ppm.	[15] 1972	journal, full paper	Dow Chemical Company	?
Vinyl chloride Rodents, inhalation	Threshold between 10 and 50 ppm.	[16] 1975	journal, prel. report	Industrial Bio-Test Laboratories	??

Table 1. (Cont'd.)

(1) Substance, Species, Route of Exposure	(2) Stokinger's Claim	(3) Reference, Year	(4) Type of Publication	(5) Institutional Affiliation	(6) Degree of Support for Stokinger's Claim
Vinyl chloride Rodents, inhalation	Threshold between 10 and 50 ppm.	[18] 1975	journal, prel. paper	Istituo di Oncologia	?? (- - -[3])
Vinyl chloride Rodents, inhalation	Biochemical threshold between 10 and 50 ppm.	[21] 1976	journal, full paper	Dow Chemical Company	?
Dimethyl sulfate Humans, inhalation	Current industrial levels are below threshold.	Pell 1972	unpublished report	du Pont Company	n.a.
Dimethyl sulfate Humans, inhalation	Current industrial levels are below threshold.	Zeller 1972	private communication[5]	BASF Company	??[1], ?[2]
Asbestos Humans, inhalation	No tumors below 125 mppgc/years.	[24, 25] 1973	journals, full papers	University of Pittsburgh	- - -

Legend for column 6:

\+ The reported results deviate significantly from the linear hypothesis.

\- The reported results are compatible with the linear hypothesis.

\- - The reported results are expected according to the linear hypothesis.

\- - - The report results provide evidence of cancer below the level that Stokinger claims to be a threshold.

? The reported data do not allow any conclusions in either direction.

?? The text referred to contains no information on the types of exposure that Stokinger claims that it describes.

n.a. The report referred to was not available in the open literature.

[1]Stokinger's reference.

[2]Stokinger's source.

[3]Later report from the same study, available but not referred to by Stokinger.

[4]Three experimental groups.

[5]Incorrectly referred to as journal article.

REFERENCES

1. H. E. Stokinger, Nonstatistical vs. Illusory Statistical Approaches to the Estimation of Risk from Environmental Chemicals, in *Dangerous Properties of Industrial Materials Report, 7,* pp. 2-8, Van Nostrand Reinhold, New York, January/February 1987.
2. W. Cook, *Occupational Exposure Limits—Worldwide,* American Industrial Hygiene Association, 1987.
3. S. O. Hansson, *Setting the Limit. Occupational Health Standards and the Limits of Science,* Oxford University Press, New York, 1998.
4. B. I. Castleman and G. E. Ziem, Corporate Influence on Threshold Limit Values, *American Journal of Industrial Medicine, 13,* pp. 531-559, 1988.
5. B. K. J. Leong, R. J. Kociba, G. C. Jersey, and P. J. Gehring, Effects from Repeated Inhalation of Parts per Billion of bis(Chloromethyl)Ether in Rats (Abstract), *Toxicology and Applied Pharmacology, 33,* p. 175, 1975.
6. B. K. J. Leong, R. J. Kociba, and G. C. Jersey, A Lifetime Study of Rats and Mice Exposed to Vapors of Bis(chloromethyl)ether, *Toxicology and Applied Pharmacology, 58,* pp. 269-281, 1981.
7. R. J. Kociba, S. B. McCollister, C. Park, T. R. Torkelson, and P. J. Gehring, 1,4-Dioxane. I. Results of 2-Year Ingestion Study in Rats, *Toxicology and Applied Pharmacology, 30,* pp. 275-286, 1974.
8. T. R. Torkelson, B. K. J. Leong, R. J. Kociba, W. A. Richter, and P. J. Gehring, 1,4-Dioxane. II. Results of a 2-Year Inhalation Study in Rats, *Toxicology and Applied Pharmacology, 30,* pp. 287-298, 1974.
9. E. Bingham, *Bioassay of Coke Oven Coal Tar Carcinogenic Activity,* National Institute for Occupational Safety and Health, 1982.
10. J. A. Zapp, Inhalation Toxicity of Hexamethylphosphoramide [Summary Report], *American Industrial Hygiene Association Journal, 36,* p. 916, 1975.
11. K. P. Lee and H. J. Trochimowicz, Induction of Nasal Tumors in Rats Exposed to Hexamethylphosphoramide by Inhalation, *Journal of the National Cancer Institute, 68,* pp. 157-164, 1982.
12. K. P. Lee and H. J. Trochimowicz, Pulmonary Response to Inhaled Hexamethylphosphoramide in Rats, *Toxicology and Applied Pharmacology, 62,* pp. 90-103, 1982.
13. K. P. Lee and H. J. Trochimowicz, Morphogenesis of Nasal Tumors in Rats Exposed to Hexamethylphosphoramide by Inhalation, *Environmental Research, 33,* pp. 106-118, 1984.
14. J. A. Zapp, HMPA: A Possible Carcinogen, *Science, 19,* p. 422, 1975.
15. C. G. Kramer and J. E. Mutchler, The Correlation of Clinical and Environmental Measurements for Workers Exposed to Vinyl Chloride, *American Industrial Hygiene Association Journal, 33,* pp. 19-30, 1972.
16. M. L. Keplinger, J. W. Goode, D. E. Gordon, and J. C. Calandra, Interim Results of Exposure of Rats, Hamsters, and Mice to Vinyl Chloride, *Annals of the New York Academy of Sciences, 246,* pp. 219-220, 1975.
17. H. E. Stokinger, The Case for Carcinogen TLVs Continues Strong, *Annals of the American Conference of Governmental Industrial Hygienists, 9,* pp. 257-264, 1984.
18. C. Maltoni and G. Lefemine, Carcinogenicity Bioassays of Vinyl Chloride: Current Results, *Annals of the New York Academy of Sciences, 246,* pp. 195-218, 1975.
19. E. Marshall, Federal Court Finds IBT Officials Guilty of Fraud, *Science, 222,* p. 488, 1983.

20. C. Maltoni, G. Lefemine, A. Ciliberti, G. Cotti, and D. Carretti, Carcinogenicity Bioassays of Vinyl Chloride Monomer: A Model of Risk Assessment on an Experimental Basis, *Environmental Health Perspectives, 41,* pp. 3-29, 1981.
21. P. G. Watanabe, R. E. Hefner, and P. J. Gehring, Vinyl Chloride-Induced Depression of Hepatic Non-Protein Sulfhydryl Content and Effects on Bromosulphalein (BSP) Clearance in Rats, *Toxicology, 6,* pp. 1-8, 1976.
22. H. Druckrey, H. Kruse, R. Preussmann, S. Ivankovic, and C. Landschütz, Cancerogene alkylierende Substanzen III. Alkyl-halogenide, -sulfate, -sulfonate und ringgespannte Heterocyclen, *Zeitschrift für Krebsforschung, 74,* pp. 241-270, 1970.
23. H. Druckrey, R. Preussmann, N. Nashed, and S. Ivankovic, Carcinogene alkylierende Substanzen I. Dimethylsulfat, carcinogene Wirkung an Ratten und wahrscheinliche Ursache von Berufskrebs, *Zeitschrift für Krebsforschung, 68,* pp. 103-111, 1966.
24. P. Enterline, P. de Coufle, and V. Henderson, Respiratory Cancer in Relation to Occupational Exposures among Retired Asbestos Workers, *British Journal of Industrial Medicine, 30,* pp. 162-166, 1973.
25. P. E. Enterline and V. Henderson, Type of Asbestos and Respiratory Cancer in the Asbestos Industry, *Archives of Environmental Health, 27,* pp. 312-317, 1973.
26. M. A. Schneiderman, Digestive System Cancer Among Persons Subjected to Occupational Inhalation of Asbestos Particles: A Literature Review with Emphasis on Dose Response, *Environmental Health Perspectives, 9,* pp. 307-311, 1974.

Note: Originally published in *New Solutions, 8*(2), pp. 175-189, 1998.

CHAPTER 2

Endocrine Disruption Comes into Regulatory Focus

Davis Baltz

Endocrine disruption has come under regulatory scrutiny since the passage of 1996 federal laws. The U.S. Environmental Protection Agency (EPA) convened the Endocrine Disruptor Screening and Testing Advisory Committee (EDSTAC) to develop recommendations for a screening and testing program to assess chemicals' potential to disrupt hormone function.

The committee's September 1998 consensus report is important because it signals that endocrine disruption is a threat which a responsible society has an obligation to address. Furthermore, the program will generate significant new toxicological data.

However, shortcomings in the EPA program will include: inadequate low dose testing; lack of a screening assay that examines early developmental exposure; and mechanisms that substitute "functionally equivalent" information.

Implementation will face serious hurdles: insufficient funding; realistic validation yardsticks; resolution of crucial scientific questions, and political will.

A wide diversity of constituencies appeared before the EDSTAC during public meetings, and the public will stay engaged by commenting on EPA's proposed program and by nominating chemicals for inclusion in the program.

A meaningful response to endocrine disruption exposes a fundamental flaw in traditional risk assessment. If there is no "safe" exposure, there is also no regulatory threshold to identify.

The publication of *Our Stolen Future* by Colborn, Dumanoski, and Myers in 1996 called public attention to an emergent issue with undeniably profound implications for public health: endocrine disruption. Increasing evidence in wildlife and in humans indicates that some synthetic chemicals interfere with normal hormone function by mimicking or blocking the action of naturally occurring hormones in the body. Endocrine disruption is implicated in a wide range of health effects from reproductive dysfunction in men and women, certain cancers, immune system impairment, and neurodevelopmental disorders [1].

Since the appearance of *Our Stolen Future,* there has been a tremendous amount of new research [2, 3] and more is on the way. Government, industry, and public interest scientists are devoting significant resources to investigating and reporting on endocrine disruption.

Much of the increased attention has been driven by provisions of the 1996 Food Quality Protection Act (FQPA) and the 1996 reauthorization of the Safe Drinking Water Act (SDWA), which require the U.S. Environmental Protection Agency (EPA) to develop a screening and testing program. The EDSTAC first met in December 1996 and continued its work until September 1998 when it presented its final report to EPA. At that time, EPA stated that it "embraced" EDSTAC recommendations, which would serve as a "blueprint" for the EPA program [4].

EPA published its proposed Endocrine Disruptor Screening Program in the Federal Register on December 28, 1998[1] to comply with the FQPA and SDWA statutes. Public comments were received until February 26, followed by scientific peer review March 30-April 1, in Washington, D.C. by a first-ever combined panel of the EPA's Science Advisory Board (SAB) and the EPA's Scientific Advisory Panel (SAP).

THE IMPORTANCE OF EDSTAC'S RECOMMENDATIONS

EDSTAC's work is an important step forward because, prompted by new federal laws, it signals that endocrine disruption is now a threat which a responsible society has an obligation to address. The EDSTAC membership, representing interests across the spectrum, agreed on consensus recommendations, reflecting this new awareness. Beyond this, EDSTAC recommendations exceeded what would be required solely by statute in some significant ways:

- EDSTAC recommended the EPA consider a much larger universe of chemicals as potential candidates for screening and testing (87,000).
- EDSTAC recommended that EPA screen and test for androgen and thyroid effects, not limiting itself solely to estrogen-related effects.
- EDSTAC recommended that EPA's screening and testing program should address environmental impacts and focus on both human and ecological health.
- EDSTAC recommended that mixtures enter the screening and testing program.

[1] http://www.epa.gov/fedrgstr/EPA-PEST/1998/December/Day-28/p34298.htm

Perhaps most important of all, the program will generate an enormous amount of new toxicological data which does not currently exist. Nearly three-quarters of high production chemicals in U.S. commerce today, for example, lack minimum data requirements for health hazard screening [6].

The new data will be crucial because actual test results inform environmental decision-making in a way that theoretical risk assessment cannot. Not since the Toxic Substances Control Act (TSCA) of 1976 has the federal government required actual testing of industrial chemicals despite the opinion of most observers that TSCA has been a dismal failure. In the twenty-plus years since TSCA provisions became effective, its authority to order testing of chemicals has been used on only 121 of them, while negotiated consent agreements and voluntary testing have been used to gather information on an additional 443 chemicals [7]. The total number of chemicals in the TSCA Inventory, by contrast, stands at 75,500 [8].

EPA'S PLAN AND ITS SHORTCOMINGS

So what will the EPA screening and testing program look like? EDSTAC recommended a tiered approach, and this conceptual framework is not controversial. After an initial *sorting* step, chemicals will undergo *priority setting* which will determine, in essence, in what order the chemicals will be screened and tested. The next step is *Tier One screening,* a first "rough cut" battery of assays, followed by *Tier Two testing,* if necessary, which would provide much more comprehensive and specific information.

Chemicals which ultimately test negative, or are not prioritized because of low concern, will be deferred until such time as new evidence indicates they should be re-examined. Chemicals which ultimately test positive as endocrine disruptors after Tier Two Testing would then undergo Hazard Assessment at EPA and subsequent regulatory action.

Having said that EDSTAC's recommendations represent a significant accomplishment, it is also the case that the EPA program will have some serious shortcomings. This is because the EDSTAC was unable to agree on all aspects of the screening and testing program and has left certain key decisions to EPA which will be taken without the consensus backing of the EDSTAC panel.

For example, perhaps the most glaring hole in the EDSTAC's recommended program is the inadequacy of low dose testing. Low dose testing issues were among the most controversial EDSTAC faced. The concern is that some chemicals seem to have effects at low doses which are not seen at higher doses [9]. If effects are not adequately addressed at all relevant doses, including current environmental "background" levels, a vital piece of the screening and testing program's mission will be missed.

Industry has downplayed the significance of existing low dose research because of the profound market implications it has for specific chemical products. During

EDSTAC deliberations, industry representatives said that in terms of hurdles to their ultimate sign-on to the EDSTAC document, the low dose issue was their "Mt. Everest" and they would not allow their products to be "held hostage to ten mice."[2] However, over the time since EDSTAC first met, there has been broader acceptance of low dose effects in the independent scientific community.[3]

Despite the increasing marginalization of industry's position, the EDSTAC was able to recommend to EPA only that a "project to address low dose issues" should be undertaken and completed within four to six months to resolve the underlying uncertainties and controversy about dose selection.

A second shortcoming of the EDSTAC recommendations is that the initial screening battery does not contain an assay that looks at early developmental exposure. Since irreversible effects from endocrine disruption may occur during these developmental life stages, the lack of a screening assay that examines prenatal or pre-hatch exposure is a compelling concern.

Third, the EDSTAC recommendations allow mechanisms for the substitution of alternative test results, or the bypassing of the screening tier altogether, if "functionally equivalent" information is available to replace data that the endocrine disruptor screening and testing program itself would generate. What exactly does "functional equivalence" mean? It raises concerns because it would usually mean that less information is used for decision-making. But if less information is utilized, can it really be "equivalent?" This is especially problematic for tests that look at reproduction studies across more than one generation.

Fourth, it will be left to EPA to generate an actual list of named chemicals to enter the program. EDSTAC recommended several criteria for EPA to use in the priority-setting process, but was unable to specify a way to weight the criteria that would yield a concrete list. EPA will have to finish this job, and no matter what the agency decides it will almost certainly anger some stakeholders, possibly leading to challenges or lawsuits that clog the system.

Fifth, there are concerns about the program's decision-making process. How will screening and testing results be interpreted? What level of proof should be required to send a chemical to the next tier of scrutiny? EDSTAC is recommending that a "weight-of-evidence" approach be utilized on what will mostly be a case-by-case basis. Many decisions may likely prove to be straightforward; the weight of evidence will clearly indicate the next path of action. But there also will be cases where equivocal evidence is in the record, and EPA decision making could be controversial.

[2] Stated in EDSTAC public meeting in Baltimore, March 17-18, 1998 by representative of Chemical Manufacturers Association.

[3] Characterizing the Effects of Endocrine Disruptors on Human Health at Environmental Exposure Levels, NIEHS, Research Triangle Park, North Carolina, May 11-13 1998.

OBSTACLES TO IMPLEMENTATION

No matter what the final details of the EPA's proposal are, its implementation faces serious hurdles. The first and most important is simply finding the funding to complete the "pre-program" work before the screening and testing program can formally begin. These key activities will revolve around the standardization and validation of proposed assays; the recommended research project to address low dose issues; and the implementation of an automated screening process called High Throughput Pre-Screening. The costs will be substantial, and public and private funds must be resourced to carry the program ahead. The standardization and validation process alone is estimated to cost at least $20 million, yet the 1999 fiscal year EPA budget has earmarked only a small fraction of this amount for the implementation of all EDSTAC recommendations. Once all the groundwork is completed and the program formally begins, presumably industry money will be used for its ongoing operation.

A second hurdle is deciding what a realistic degree of validation should be. The proposed assays must undergo standardization and validation, but it is important that the bar not be set so high that progress is delayed for years or longer. There is ample potential for foot-dragging delays if industry does not make good-faith efforts to move forward on the validation process. Yet some observers would say that for industry, to delay is to win.

A third key task will be the resolution of crucial questions of scientific evidence. Industry representatives on EDSTAC blocked low dose testing and the inclusion of a screening assay for prenatal exposure because they claimed the research was inconclusive. EPA must aggressively move to resolve this issue and then compel low dose testing and developmental exposure assays in the screening tier if the evidence warrants.

A fourth obstacle will be whether or not EPA can summon the political will in the face of immense pressure from industry to both carry forward the screening and testing program and then to make meaningful regulatory decisions that reduce exposure to endocrine disruptors. Activities to date have been shepherded by Dr. Lynn Goldman, EPA Assistant Administrator in the Office of Prevention, Pesticides and Toxic Substances, who also served as EDSTAC's Chair. Dr. Goldman deserved much of the credit for EDSTAC's important consensus report, but she resigned from the EPA at the end of 1998. It will be important to find a committed and savvy replacement, but with less than two years left in the Clinton Administration, it will be difficult to expect the new Assistant Administrator to serve in anything other than an "acting" capacity, diminishing his or her influence in carrying forward the endocrine disruptor screening and testing program.

THE PUBLIC'S ROLE

The public's concern has grown steadily as endocrine disruption has burst into public awareness. The 1998 national broadcast of the issue on the PBS

"Frontline" program ("Fooling With Nature," which first aired on some affiliates on June 2, 1998) is testament to its high profile and high stakes.

During its life, the EDSTAC held public meetings in seven U.S. cities. A remarkable diversity of constituencies appeared before the committee to provide public comments. Not only environmental groups and industry spoke. EDSTAC also heard from those who identified themselves as endometriosis associations, the chemically injured, cancer survivors, learning disability networks, environmental justice advocates, consumers, mothers, nursing mothers, grandmothers, pregnant women, doctors, nurses, public health professionals, farmers including organic growers, ranchers, farmworkers, gardeners, unions (Oil, Chemical, and Atomic Workers; Southwest Public Workers Union; Farmworkers Association of Florida), small businesses, environmental consultants, teachers, students, the PTA, the League of Women Voters, conservationists, veterans, birth defect researchers, DES daughters, and "ordinary citizens."

The public played an extremely important role by consistently demonstrating to EDSTAC that they were affected and they cared about keeping their communities and families safe. The public will stay engaged because every single person in the world today carries a body burden of endocrine disrupting chemicals, yet no one has granted permission for this chemical trespass into the ultimate privacy of our own bodies. This simple fact makes everyone a stakeholder.

In addition to contributing public comments to EPA's proposed program, citizens also will be able to nominate chemicals for inclusion in the screening and testing program. The primary objective will be to capture chemicals that are not identified by other means during priority setting, but are of concern to communities or populations that might have disproportionately high exposures.

RISK ASSESSMENT UNDER FIRE

Debate about how to meaningfully respond to the threat of endocrine disruption has exposed a fundamental flaw in traditional risk assessment. Traditional toxicology has conducted testing to identify the exposure level at which no effect is seen, a "no observed adverse effect level" (NOAEL). The assumption has been that exposure below this point is essentially safe, and so, after building in a safety margin, there is no need to regulate a substance if exposure levels fall below the NOAEL.

However, endocrine disruption low dose research is telling us that adverse effects (and effects whose impacts cannot yet be determined) are occurring at levels which are orders of magnitude below the identified NOAEL. In other words, there is no "safe" exposure. And if there is no acceptable level of exposure, the sensible and precautionary approach would dictate that the chemical be sunsetted.

This is why industry is so alarmed about endocrine disruption. Industry will not be able to muster a defense couched in the familiar terms of "sound science."

Certain chemicals, and the consumer products that are made from them, will be identified as endocrine disruptors and regulated or banned. One has only to reflect for a moment on the sectors of the economy that rely on chemicals to see how profound an impact this could have.

For now, industry continues to stonewall by denying that scientific evidence exists. But as the research evidence builds, industry's continued denials will erode its credibility.

If endocrine disruption finally forces us to adopt a new paradigm for examining and managing risk, it will usher in a new set of rules about how we organize ourselves as a society. If we are truly threatening our fertility, intelligence, and survival, then we must change our behavior.

REFERENCES

1. T. Colborn, D. Dumanoski, and J. P. Myers, *Our Stolen Future,* Dutton Press, New York, 1996.
2. See World Wildlife Fund's issue brief, *Chemicals that Compromise Life: A Call to Action,* 9/98, for a review of recent findings.
3. *Toxicology and Industrial Health, 14*:1-2, January-April 1998, devotes an entire issue to endocrine disruption, with a primary focus on neurodevelopmental effects.
4. U.S. EPA, Endocrine Disruptor Screening and Testing Advisory Committee, Final Report, Washington, D.C., August 1998.
5. 5 October 1998 Washington, D.C., briefing by Dr. Lynn Goldman, EPA Assistant Administrator in the Office of Prevention, Pesticides and Toxic Substances and EDSTAC Chair.
6. *Toxic Ignorance: The Continuing Absence of Basic Health Testing for Top-Selling Chemicals in the United States,* Environmental Defense Fund, New York, p. 15, 1997.
7. U.S. EPA, *Endocrine Disruptor Screening and Testing Advisory Committee,* Final Report, Washington, D.C., pp. 2-11, August 1998.
8. U.S. EPA, *Endocrine Disruptor Screening and Testing Advisory Committee,* Final Report, Washington, D.C., pp. 2-10, August 1998.
9. F. S. vom Saal, B. G. Timms, M. M. Montano, P. Palanza, K. A. Thayer, S. C. Nagel, M. D. Dhar, V. K. Ganjam, S. Parmingiani, and W. V. Welshons, Prostate Enlargement in Mice Due to Fetal Exposure to Low Doses of Estradiol or Diethylstilbestrol and Opposite Effects at High Doses, *Proceedings of the National Academy of Science, 94*, pp. 2056-2061, 1997.

Note: Originally published in *New Solutions, 9*(1), pp. 29-35, 1999.

CHAPTER 3

The Relevance of Occupational Epidemiology to Radiation Protection Standards

Steve Wing, David Richardson, and Alice Stewart

Large-scale epidemiological studies of U.S. Department of Energy workers have been underway since the 1960s. Despite the increasing availability of information about long-term follow-up of badge-monitored nuclear workers, standard-setting bodies continue to rely on the Life Span Study (LSS) of A-bomb survivors as the primary epidemiological basis for making judgments about hazards of low-level radiation. Additionally, faith in the internal and external validity of studies of A-bomb survivors has influenced decisions about the design, analysis, and interpretation of many worker studies. A systematic comparison of the LSS and worker studies in terms of population characteristics, types of radiation exposures, selection factors, and dosimetry errors suggests that the priority given to dose response findings from the LSS is no longer warranted. Evidence from worker studies suggests that excess radiation-related cancer deaths occur at doses below the current occupational limits; low-dose effects have also been seen in studies of childhood cancers in relation to fetal irradiation. These findings should be considered in revising current radiation protection standards.

Environmental and occupational standards are based as much on policy and advocacy as on science; however, scientific studies play a role in justifying standards [1-4]. Studies of cells in vitro and in vivo, animal experiments, and studies of molecular markers of disease may contribute to the scientific evidence. However, epidemiological studies of disease in human populations are especially important in risk estimation and standard-setting because other approaches necessitate extrapolation from high to low doses, from molecules and cells to organisms, and from other species to humans [5-7].

The ability of external penetrating ionizing radiation (either gamma radiation or X-rays) to cause biological damage was first recognized shortly after the discovery of x-rays in 1896 [8]. A radiation protection standard for workers was first introduced in 1925, and the maximal permissible dose was reduced many times over the next century (Figure 1). The first standards were based on the ability of ionizing radiation to produce acute effects such as erythema (skin reddening) that could be observed immediately in small numbers of exposed persons. However, the appearance of cancers among radiation workers and nuclear physicists showed that radiation could produce neoplastic effects and led to the recognition of the importance of long-term epidemiological studies for understanding radiation risks and setting standards. The current protection standard used by the International Commission on Radiation Protection is 20 mSv whole body annual dose equivalent (Figure 1) [6], however, the United States still permits workers to receive up to 50 mSv per year.

Epidemiological studies have been conducted of medical, environmental, occupational, and military exposures to radiation [5]. Unlike experimental research, these studies lack direct control over measurement and exposure conditions. Radiation doses are often not recorded and may be difficult to estimate; the people included in studies may constitute unusual populations, such as patients with particular diseases, whose relevance to other populations is not clear; and evaluation of many possible outcomes, especially those that are nonfatal, may be difficult or impossible. Furthermore, studies of long-term effects of exposure may suffer from problems of loss to follow-up; and, within a single study population, people with lower and higher doses may differ with respect to confounding factors that affect disease rates even in the absence of radiation exposure. The reliability of the evidence from epidemiological studies therefore depends on their ability to be designed and analyzed to reduce the influence of such biases on the results.

Although workers were exposed to ionizing radiation in various industries prior to World War II, a totally new exposure situation was created in the United States by the Manhattan Project for developing nuclear weapons. Hundreds of thousands of workers were recruited to a new industry situated in remote and secret locations [9]. Because some hazards were recognized in advance, workers at many of these facilities were monitored for radiation exposures, and a new professional discipline, health physics, was created to provide radiological protection (Figure 2) [10, 11]. While the immediate goal of health physicists was worker protection, systematic collection of dosimetry records created an opportunity to investigate relationships between repeated exposure to small doses of radiation and disease that would be relevant to future standard setting.

However, the most influential epidemiological study of ionizing radiation today is not based on follow-up of badge-monitored workers, but on follow-up of

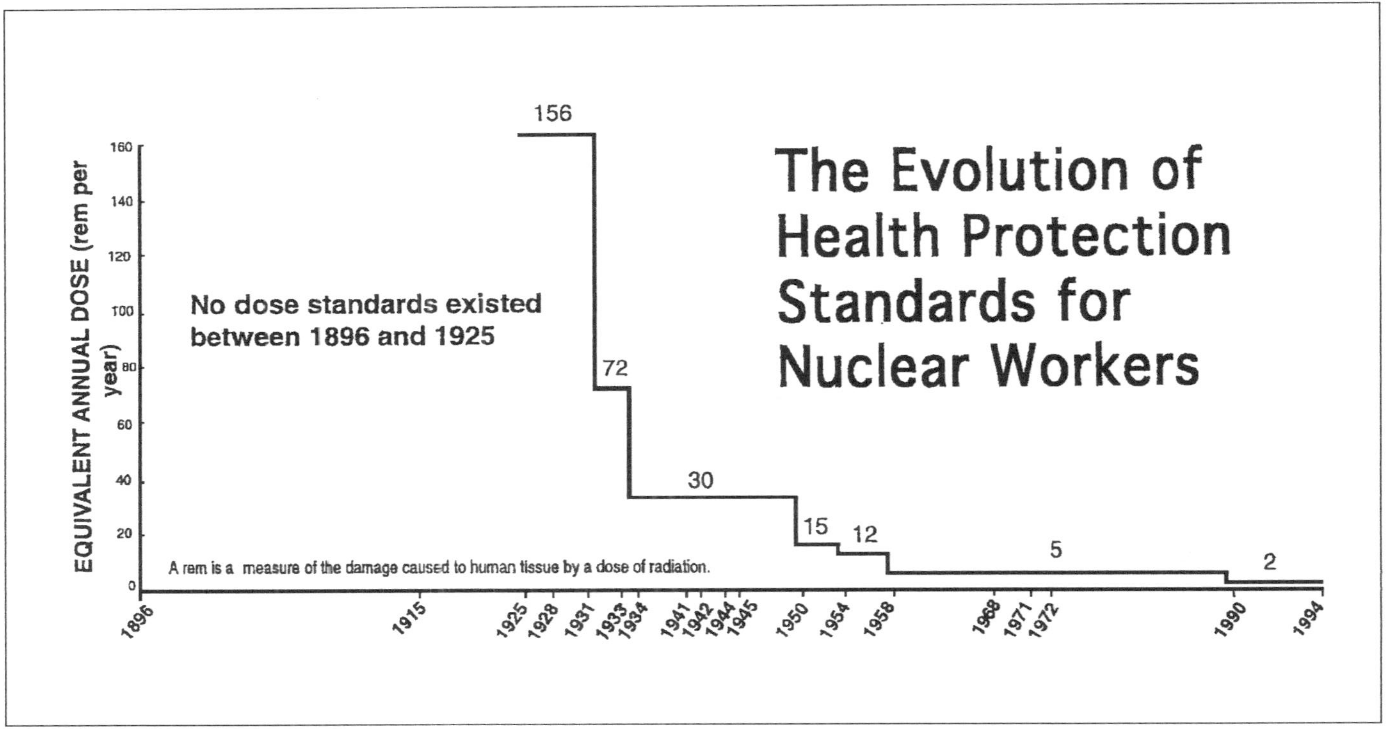

Figure 1. The evolution of health protection standards for nuclear workers. **Source**: *Closing the Circle on the Splitting of the Atom/The Environmental Legacy of Nuclear Weapons Production in the United States and What the Department of Energy is Doing About It*, U.S. Department of Energy, Office of Environmental Management, January 1995.

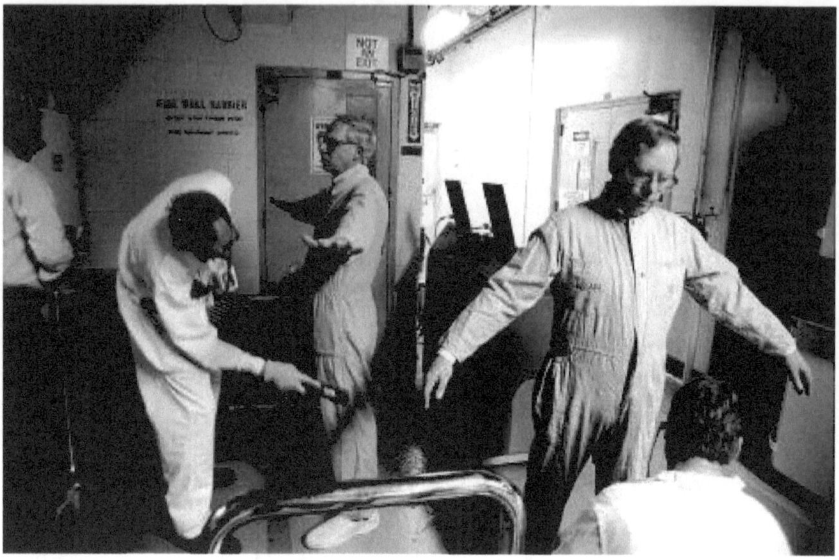

Figure 2. Workers are checked for contamination by radiation at the nuclear weapons plant at Rocky Flats, Colorado. Photo by Robert Del Tredici.

the people who were affected by the bombs produced during the Manhattan project [5]. By the 1960s, when the first large-scale study of nuclear workers was finally initiated more than two decades after occupational dose records began to be collected, many scientific papers and technical reports about Japanese A-bomb survivors had already been written. These reports were widely interpreted as showing that the effects of occupational exposures to radiation would be too small to be detected in epidemiological studies of U.S. nuclear workers [12-14]. However, questions about the reliability of A-bomb findings already had begun to emerge due to the discrepancy between findings of excess risks of childhood cancer following medical irradiation in utero and the absence of an effect among A-bomb survivors who had been exposed in utero [15, 16].

During the last two decades numerous studies of nuclear workers have suggested that radiation risk estimates based on A-bomb survivors could be substantially underestimating the cancer risks from protracted low-level exposure to radiation [4, 17-26]. However, the worker studies have not led to revision of radiation protection standards; rather, the A-bomb survivor study remains the principal source of cancer risk estimates and radiation safety standards, as well as the principal referent for occupational radiation researchers who consider the A-bomb findings to be a standard for interpreting nuclear worker findings [27-31]. The main purpose of this chapter is to reappraise the relevance of nuclear worker and A-bomb survivor studies to radiation safety standards. We also review aspects of the scientific climate that may contribute to the dominance of studies of A-bomb survivors, and consider reasons that studies of nuclear workers should receive more attention as a basis for understanding the health effects of repeated exposure to small doses of ionizing radiation.

COMPARISON OF WORKER AND A-BOMB SURVIVORS STUDIES

Japanese doctors began to observe radiation effects even as they were caring for survivors of the nuclear attacks on Hiroshima and Nagasaki [32]. The first major effort to conduct epidemiological studies, however, began during the occupation of Japan by United States forces after World War II when the U.S. government established the Atomic Bomb Casualty Commission (ABCC) under the control of the Supreme Allied Commander. Although Japanese doctors and nurses were necessarily involved in collecting data, the studies were designed and run by scientists from the United States [32], and funding for the successor to ABCC, the Radiation Effects Research Foundation (RERF) continues to come from the United States as well as Japan.

Three types of studies among A-bomb survivors can be distinguished: studies of inherited genetic effects among children whose parents had been exposed to A-bomb radiation; studies of various effects of fetal irradiation; and studies of long-term effects of postnatal exposure to radiation. This last study, called the Life Span Study (LSS), has primarily addressed cancer as an outcome and is the main source of cancer risk coefficients that are widely used to estimate the impact of radiation on cancer among workers and other exposed populations.

The possibility of conducting large-scale epidemiological studies of workers in the nuclear industry was discussed by the Atomic Energy Commission's Advisory Committee for Biology and Medicine in the early 1950s. However, despite the existence of extensive programs to monitor worker exposure to radiation, as well as occupational medicine programs that could have documented biological effects, epidemiological studies were not begun until the 1960s. At that time, Thomas Mancuso of the University of Pittsburgh, a physician who had extensive experience in occupational epidemiology, was asked to direct the U.S. worker studies. Surveys of nuclear workers also have been conducted in other countries (notably the U.K.) and there has been pooling of worker studies from the United States, the United Kingdom, and Canada [22, 29, 30].

When comparing the findings of epidemiological studies, it is not sufficient to consider the results of each study equally; rather, it is critical to evaluate the quality of the study information and how the data were analyzed. Some of the design issues relevant for consideration when comparing the A-bomb survivor study with studies of nuclear workers are summarized in Table 1.

Table 1. Epidemiological Basis for Evaluating External Radiation and Cancer: A-Bomb and Worker Studies

Design Issue	Life Span Study	Worker Studies
Study population	Survivors in 2 cities; all ages	Employee rosters from nuclear facilities; adults
Sample size	75,000-120,000	A few thousand (single facility) to hundreds of thousands (combined facilities)
Exposure	Acute gamma and neutron doses, low to lethal	Chronic low doses, primarily gamma
Exposure measurement	Physical models, survey responses	Individual dosimeters
Outcomes	Cause-specific mortality; cancer incidence; medical surveillance	Cause-specific mortality
Selection factors	Selective survival, chronic effects of acute exposure	Healthy worker effects: – worker selection – worker survival

Study Population

A well-defined study population is relevant to evaluating the generalizability of a study, comparability between exposure groups, and factors related to the susceptibility of a population to the exposure of interest.

Although studies of A-bomb survivors began shortly after the bombings, a list of persons for inclusion in the LSS cohorts was not obtained until after the 1950 census, when individuals were interviewed to collect information about location at the time of the bombing, exposure position, shielding, and acute radiation effects. The people included in subsequent studies have changed many times [33], but in each cohort there have been persons of all ages and a wide range of estimated doses for persons at increasing distances from the hypocenter.

In contrast, studies of nuclear workers in the United States are based on employee rosters taken from primary contractors of the Department of Energy (DOE). In general, prime contractors kept careful records of employees for payroll and security purposes. Epidemiological studies, therefore, begin with an essentially complete list of workers who have ever been employed in a brand new industry. As in other occupational studies, worker exposures generally occur between sixteen and seventy years of age.

Sample Size

The LSS study is exceptionally large, and the survivors received a wide range of radiation doses. The cohort with the latest (DS86) dose estimates includes 75,991 survivors, as described in the 1990 report of the National Academy of Science Committee on the Biological Effects of Ionizing Radiation (BEIR V) [5]. Large sample size is a major strength of any survey and contributes to the statistical precision of any radiation risk estimates.

Worker studies vary in size depending upon the number of employees at a facility and eligibility criteria for entry into the study, such as race, gender, date of hire, length of employment, and availability of radiation exposure records. Many DOE worker studies are large, and some of the facilities had tens of thousands of badge-monitored workers on their rosters. Although sometimes workers from several facilities have been combined to increase the study size, this may introduce problems due to differences in the types of exposures, measurements, and selection factors [22, 34].

Exposures

Exposure situations were very different for A-bomb survivors and workers. The whole body exposures of A-bomb survivors came primarily from gamma and neutron radiation emitted during the first microseconds of atomic explosions. Estimated doses ranged from low at long distances from the hypocenter to lethal near ground zero. The high doses of some survivors have been considered a major strength of the A-bomb studies. The RERF studies have ignored doses from irradiated structures and fallout, although Japanese doctors, veterans, and other observers have suggested that these subsidiary exposures might have been important [33, 35, 36]. Fallout that occurred at some distance from the hypocenter of the blasts would be a particular concern for epidemiological studies because it would lead to systematic underestimation of doses for survivors with lower doses from the flash [37].

In contrast, radiation exposures received by nuclear workers occur in research, production of radioactive materials (see photograph), power generation, waste disposal, and assembly of weapons. With the exception of unusual events such as criticality accidents, worker dose rates are too low to produce acute symptoms. The exposure of major interest in many worker populations is cumulative whole body exposure to penetrating ionizing radiation. In facilities that process radionuclides, internal radiation contamination may also be important. Compared to the exposures received by A-bomb survivors, nuclear workers usually receive many small doses spread over a long period of time. It is this pattern of exposure which is of primary concern for radiation protection standards.

Exposure Measurement

Quantification of a relationship between exposure and disease hinges on the ability to differentiate between people with different doses. Studies of the A-bomb survivors began by using distance from ground zero as a surrogate of radiation dose [36], but later methods of dose estimation also took account of shielding and other factors, and major revisions were announced in 1965 (T65 doses) and again in 1986 (DS86 doses). These estimates depend on the quality of two pieces of information: the amount and types of radiation emitted from the bombs, and the location of the person at the time of the bombing. Estimates of the types of radiation emitted came from models of the bombs themselves, from measurements made on test weapons, and from the physical state of the devastated cities. Revised estimates of the relative contribution of neutron and gamma radiation produced by the bombs were the main reason for changes in dose estimates in 1986 [5]. These changes, which led to substantial changes in risk estimates, illustrate the problem of using indirect estimates of dose.

The accuracy of information about the physical location and shielding of each survivor at the time of the blast depended upon the memory of the respondents and the ability of the interviewers to elicit valid information. For A-bomb survivors, problems were introduced by strained relationships between Japanese survivors and the American-run research teams, the effects of a profound disaster on memory, and the social stigma attached to being an "exposed person." Thus, in Japan, eligibility for an arranged marriage could be hampered by concerns about genetic effects of radiation, and some survivors did their best to prevent even their spouses from knowing that they had been exposed [32]. Evidence of acute injuries such as hair loss, erythema, and chromosome damage has been found among survivors whose supposedly low doses could be the result of false reporting of exposure positions [39-40].

Estimates of worker doses are based both on dosimeters incorporated into the security badges worn by nuclear workers (external radiation) and on urine tests (internal radiation). At Hanford and Oak Ridge, annual dose estimates based on the dosimeter readings are available for most of the workers included in epidemiological studies but are sometimes missing for certain years [41, 42]. Furthermore, workers may fail to wear their badges at all times, errors are made in matching hundreds of thousands of dosimetry readings to workers, and changes in dose estimates may occur due to changes in the types of dosimeters used and in the conduct of the dosimetry programs. Despite these problems, the availability of individual physical measurements of radiation exposures for workers is a unique advantage of epidemiological studies in this industry.

Outcomes

The LSS and worker studies have depended primarily upon death certificate data to identify outcomes of special interest, particularly cancers. Despite the limited medical information available on death certificates, broad categories of cancer are well-recorded, and cancer diagnoses are more accurate than diagnoses for many other causes of death [43]. Additionally, cancer registries in Hiroshima and Nagasaki were used to measure incidence of cancer, including nonfatal cases, and the LSS included medical surveillance of many survivors. Most worker studies have been limited to cause-specific mortality because of lack of reliable information about nonfatal illnesses. Lack of information on nonfatal cancers and nonmalignant diseases limits the ability of epidemiological studies to identify many problems of importance to public health, risk assessment, and standard setting.

Confounding or Selection Factors

The validity of dose response coefficients depends upon the degree to which higher and lower dose groups are similar with respect to other risk factors for the outcome of interest. This may be achieved through design, as in experimental control, or through analysis, as in adjustment for confounding factors.

Most researchers have assumed that A-bomb survivors near ground zero were not systematically different from survivors who were farther away and that, as a consequence, the LSS would be a "natural experiment." However, survival during the five years between the bombings and the beginning of the LSS, a period during which public health infrastructures of medical care, sanitation, water, and food supply were disrupted, would have favored hardy, robust persons. In fact, there is a deficit of higher dose survivors among children and older adults compared to middle ages, which suggests selective survival was most pronounced at the most biologically vulnerable ages [44]. Stewart and Kneale have presented evidence that dose-related selective survival of persons less sensitive to subsequent carcinogenic effects of radiation has led to a downward bias in estimates of radiation risks in the LSS [44-46]. However, continued use of the LSS for cancer risk estimation shows that many scientists continue to believe that "A-bomb survivors, apart from their radiation dose, are representative human beings" [47].

Studies of nuclear workers must also address bias from selection and confounding. Many studies compare cancer rates among nuclear workers to cancer rates among the general population. However, people who are employed (particularly in well-paid jobs that require high levels of education) are, in general, healthier than average, producing a healthy-worker selection effect [48-50]. Furthermore, occupational exposures to radiation are spread over several decades, and, to reach higher dose groups, workers must remain in reasonably good health for many years. Thus, workers who are more prone to get cancer may be less well-represented in higher dose groups because they leave employment or are reassigned to jobs with less exposure, producing a healthy-worker survivor effect [51]. Other differences between more and less heavily exposed workers may be created by assignment of well-trained persons with lower risk profiles to areas where there are exposures to external radiation as well as other hazards. As in the A-bomb survivor studies, problems of selection and confounding may obscure evidence of relationships between radiation and cancer; however, careful analysis can address many issues of selection and confounding.

Summary

Neither the worker studies nor the A-bomb survivors studies are free of measurement problems and other biases. Nevertheless, studies of A-bomb survivors have continued to play a predominant role in radiation risk estimation despite their focus on an extreme exposure situation, evidence of selective survival, and unresolved questions about inaccuracy of dose estimates. Occupational studies, which investigate low-level exposures similar to those of regulatory concern and have advantages of individual dose measurements, lack of reliance on interviews, and absence of selection related to surviving an atomic attack (Table 1), have been kept in the background.

INFLUENCE OF THE A-BOMB STUDIES ON STUDIES OF NUCLEAR WORKERS

Worker studies have been justified on grounds that they provide a check on the more reliable studies of A-bomb survivors, and assessment of the validity of findings of occupational studies have been based on their agreement with findings from A-bomb studies. This practice is based on the assumption that extrapolating low-dose risk from the studies of high doses is superior to estimating risk at low doses directly because inferences from the A-bomb survivor studies lead to the prediction that worker exposures are usually too low to produce detectable effects [12, 13].

Table 2 presents quotations from the recent occupational epidemiology literature that illustrate these perspectives. The first four quotations, drawn from studies of international, British, and U.S. worker populations, indicate that occupational studies are typically compared to the A-bomb studies as a gold standard [28-30]. They also indicate the role of the LSS in standard-setting. Most importantly, where evidence from worker studies disagrees with studies of the A-bomb survivors, researchers have assumed that worker studies, rather than studies of A-bomb survivors, are in error. This attitude is evident in the second quotation from Gilbert et al. (Table 2). Rather than considering the possibility that the differences in age-related sensitivity to radiation between worker and A-bomb studies might be due to differences in the exposure situations or selective survival among A-bomb survivors, they assume that differences are due to unnamed "biases in the data that are not well understood" [30].

Table 2. The Influence of Studies of A-Bomb Survivors on Studies of Nuclear Workers

Citation	Quotation
Cardis et al., 1995	A primary objective of studies of cancer risk among nuclear industry workers is the assessment of the adequacy of existing protection standards . . . based on risk estimates derived from analyses of the mortality of atomic bomb survivors and studies of other high dose exposures.
Kendall et al., 1992	One of the objectives of studies of radiation workers is to obtain direct estimates of risks from exposures to low doses of radiation at low dose rates, for comparison with . . . high dose and high dose rate exposures of the Japanese atomic bomb survivors. . . .
Kendall et al., 1992	The greater dose range and longer follow-up of the Japanese atomic bomb survivors give that study greater statistical power than the National Registry for Radiation Workers.
Gilbert et al., 1993	The analyses expressing risks and confidence limits as multiples of BEIR V predictions (Table VII), which allowed for variations in risk by cancer type, time since exposure, an age at exposure, provide stronger confirmation that risks have not been underestimated by high-dose extrapolation. . . .
Gilbert et al., 1993	If, as seems more likely, the age effects result from biases in the data that are not well-understood, this raises questions about the general validity of using these data for evaluating risks at low doses and dose rates, and certainly suggests caution in interpreting results from studies of nuclear workers.
Sever, 1988	This model was chosen because linear models have commonly been used in quantifying genetic effects due to radiation; in particular, estimates from the Japanese A-bomb survivors are based on a linear model.
Neel, 1994	The genetic data of greatest relevance to the interpretation of the findings of the Gardner-group studies are those collected in the aftermath of the Hiroshima-Nagasaki bombings.
Sever, 1988	Probably the most important consideration in interpreting these results is the contradictory evidence from other studies, particularly the Japanese A-bomb survivor studies.

The first quotation from Sever indicates that statistical models chosen for studies of workers are justified on grounds that they were appropriate for A-bomb survivors. Both Neel and Sever, in considering the evidence regarding the potential for occupational radiation exposures of workers to influence cancer among their offspring, considered the contradictory evidence from A-bomb survivor studies to be key to their conclusions that evidence of such a relationship from worker studies should be rejected (Table 2) [27, 31].

Discrepancies between worker studies and A-bomb studies also have been ignored. An increasing number of worker studies suggest both that there is excess risk of cancer at doses well below the current occupational standards, and, contrary to the LSS, that sensitivity to carcinogenic effects of radiation increases with age among adults [22-24, 26]. However, rather than consider the possibility that the LSS is not representative of the original exposed populations, researchers have assumed that worker studies are biased (Table 2).

Stewart and colleagues first questioned the general validity of A-bomb data after observing increases in childhood cancer in relation to in utero exposure to radiation and noting that no similar effects were observed in A-bomb data [52, 53]. Subsequently, risks from in utero exposure were demonstrated in other populations [54, 55]. Although the practice of obstetric radiography was greatly curtailed, this evidence that A-bomb data might be an unreliable source of information about low dose radiation risks was ignored [5]. As in the case of nuclear worker studies (Table 2), a primary reason given by the BEIR V committee for discounting evidence from studies of obstetric radiation was the fact that there was no similar finding among A-bomb survivors.

THE CULTURAL CONTEXT OF RADIATION EPIDEMIOLOGY

Many historians and philosophers of science as well as practicing scientists have noted that the content of scientific knowledge is affected not only by theories and methods of observation and experiment but by cultural forces from both inside and outside the scientific community [4, 8, 56, 57]. The weight given to evidence from the A-bomb survivor studies has some basis in methodological considerations such as the LSS's large size and range of dose variation. However, such considerations cannot explain the deference researchers and policymakers show for this study over other epidemiological investigations, including worker studies, that have important methodological advantages over the A-bomb studies. This situation suggests we must look further for explanations of the dominance of studies of A-bomb survivors in the radiation epidemiology literature.

Almost all nuclear research in the United States has been under the control of the government and large military contractors [37]. Thus, organizations with political and economic interests in promoting nuclear weapons and industrial uses of nuclear technology have also been responsible for research on the health effects of the technology [58]. The climate of secrecy surrounding military research extended into biology and medicine [59]. Recently declassified documents uncovered by the President's Advisory Committee on Human Radiation Experimentation are indicative of the influence of concerns about liability and public reaction on biomedical research and radiation protection [60]. A 1948 Atomic Energy Commission memo regarding studies suggesting that occupational standards might be too high stated:

> We can see the possibility of a shattering effect on the morale of the employees if they became aware that there was substantial reason to question the standards of safety under which they are working. In the hands of labor unions the results of this study would add substance to demands for extra-hazardous pay. . . knowledge of the results of this study might increase the number of claims of occupational injury due to radiation and place a powerful weapon in the hands of a plaintiff's attorney [60, p. 629].

A 1947 AEC memo from the Oak Ridge Medical Advisor's Office stated:

> Papers referring to levels of soil and water contamination surrounding Atomic Energy Commission installations, idle speculation on the future genetic effects of radiation and papers dealing with potential process hazards to employees are definitely prejudicial to the best interests of the government. Every such release is reflected in an increase in insurance claims, increased difficulty in labor relations and adverse public sentiment [60, pp. 626-627].

These attitudes prevailed during the period when studies of A-bomb survivors were beginning. Early scientific concerns that the studies of survivors might be useless due to problems with data collection and measurement were forgotten as reports appeared showing little evidence of genetic damage from radiation [32]. This was a relief to authorities who were seeking to expand nuclear weapons and power programs. Speculation about an end to the A-bomb studies was replaced by additional funding and an emerging consensus about the importance of studies of this unique population for understanding radiation health effects in humans [32]. Professional careers and a scientific literature were built upon studies of the survivors.

The evolving consensus about radiation risk estimates based on the A-bomb studies suggested, by the early 1960s, that dose response relationships between radiation and cancer would not be detectable in nuclear worker populations. Sterling argued that it was this very consensus which led, in part, to the initiation of worker studies in 1964 [1]. W. M. Schull, a central figure in the A-bomb research, commented on the prospects of occupational epidemiology:

> It seems to be highly probable that if one went through the mechanics of calculating the kinds of radiation effects which a study of the present magnitude might detect, one would be led to conclude that the undertaking is

> a hopeless one. However, as earlier recognized, it may have other merit in that it may provide a firmer basis for settlement of claims against the Atomic Energy Commission [1, p. 40].

S. Marks, who served as Mancuso's project officer for the nuclear worker studies, wrote:

> This study probably will not confirm or refute any important hypothesis but should permit a statement to the effect that a careful study of workers in the industry has disclosed no harmful effects of radiation (if the results are negative as they are likely to be). That statement, supported by appropriate documentation, would seem to justify the existence of the study [1, p. 41].

As a consultant to the AEC, B. MacMahon of Harvard stated:

> In my opinion this study does not have, never did have, and never (in any practical sense) will have, any possibility of contributing to knowledge of radiation effects in man . . . I recognize that much of the motivation for starting this study arose from the "political" need for assurance that AEC employees are not suffering harmful effect [1, p. 40].

This political need was satisfied for over a decade while the effort to assemble worker records on employment, radiation doses, and causes of death proceeded at a pace determined, in part, by low levels of support in comparison to funding for the A-bomb efforts. The first worker study in a scientific journal appeared in 1977, when Mancuso, Stewart, and Kneale reported an association between radiation exposures and cancer death among Hanford workers [17].

Mancuso's contract was terminated in 1978 [61], and responsibility for epidemiological studies of nuclear workers was transferred to groups at Oak Ridge, Los Alamos, and Hanford that had close ties to DOE and the scientific culture built around the A-bomb survivor studies. Gregg Wilkinson, principal investigator of the plutonium worker studies at Los Alamos National Laboratory from 1980-87, wrote that the scientific consensus at DOE was that "exposures were too low to be responsible for the illnesses that were being experienced" by workers [62]. In response to his 1987 report of increased cancer risks among Rocky Flats workers, Wilkinson was admonished by his superiors that he should "write to please the DOE because they provided funding and support" [62]. An independent review of DOE's epidemiology program found that "Secrecy has plagued the entire operation and is totally inappropriate in investigations of health and safety" [58].

Although responsibility for epidemiological studies of nuclear workers was transferred subsequently to the National Institute for Occupational Safety and Health (NIOSH), data-access problems remain and are routinely discussed at NIOSH research and advisory committee meetings. Data access, open scientific inquiry, and adequate funding for worker studies are all clearly important.

However, these are in some ways minor issues in comparison to the problem of the momentum established during a half-century of epidemiological research dominated by studies of A-bomb survivors. Even with open data access and adequate funding, researchers investigating radiation health effects among nuclear workers will have to overcome the constraints imposed by this scientific culture upon hypothesis generation, design, analysis, and interpretation of occupational studies.

CONCLUSION

For more than a half century the study of radiation health effects has been strongly influenced by military and industrial interests. These interests are so deeply woven into the fabric of the scientific culture that scientists participating in the dominant research programs may be unaware of the impact that secrecy, deference to authority, and disdain for concerns of workers and the public has had on the professional culture in which their research takes shape.

We have argued that one reflection of this scientific culture is an increasingly outdated emphasis on evidence about radiation health effects based on studies of A-bomb survivors, a select population whose doses from one-time flash exposures were not measured. Another reflection of this culture is that the A-bomb studies have been used as a lens through which studies of radiation-exposed worker populations are viewed. Thus, despite the unique opportunity to conduct epidemiological studies of low doses and low-dose-rate exposures that have been quantified with individual dosimeters, occupational studies have been viewed as merely a check on findings from A-bomb studies. This view has contributed to delayed attention to studies of worker health, inadequate resources for collection of medical and exposure data, and a failure to fully develop sensitive analytical approaches appropriate for epidemiological studies of workers in their own right.

Scientific attention to nuclear worker studies should increase in the future. Longer follow-up and larger numbers of deaths will increase their statistical power and opportunities for analysis of rare causes of death, disease latency, influences of age at exposure, and other aspects of susceptibility. Greater attention to historical records at DOE facilities should allow better measurement of radiological and other exposures. As researchers and policymakers come to appreciate the unique advantages of studies of nuclear workers, these studies should make a greater impact on occupational and environmental exposure standards.

ACKNOWLEDGMENT

Wing and Richardson's work was supported in part by grant 5 R01 CCR412931-02 from the National Institute for Occupational Safety and Health.

REFERENCES

1. T. D. Sterling, The Health Effects of Low-Dose Radiation on Atomic Workers: A Case Study of Employer-Directed Research, *International Journal of Health Services, 10,* pp. 37-46, 1980.
2. J. C. Robinson and D. G. Paxman, Technological, Economic, and Political Feasibility in OSHA's Air Contaminants Standard, *Journal of Health Politics, Policy & Law, 16,* pp. 1-18, 1991.
3. S. M. Rappaport, Threshold Limit Values, Permissible Exposure Limits, and Feasibility: The Bases for Exposure Limits in the United States, *American Journal of Industrial Medicine, 23,* pp. 683-694, 1993.
4. R. H. Nussbaum, The Linear No-Threshold Dose-Effect Relation: Is It Relevant to Radiation Protection? *Medical Physics, 25,* pp. 291-299, 1988.
5. National Research Council, Committee on the Biological Effects of Ionizing Radiation, *Health Effects of Exposure to Low Levels of Ionizing Radiation (BEIR V),* National Academy Press, Washington, D.C., 1990.
6. International Commission on Radiological Protection, *1990 Recommendations of the International Commission on Radiological Protection,* Vol. 21, Oxford, Pergamon Press, 1991.
7. United Nations Scientific Committee on the Effects of Atomic Radiation, *Sources and Effects of Ionizing Radiation,* United Nations, New York, 1993.
8. C. Caufield, *Multiple Exposures: Chronicles of the Radiation Age,* University of Chicago Press, Chicago, 1989.
9. R. Rhodes, *The Making of the Atomic Bomb,* Simon & Schuster, New York, 1986.
10. H. D. Smyth, *Atomic Energy for Military Purposes: The Official Report on the Development of the Atomic Bomb under the Auspices of the United States Government, 1940-1945,* Princeton University Press, New Jersey, 1945.
11. K. Z. Morgan, Health Physics: Its Development, Successes, Failures, and Eccentricities, *American Journal of Industrial Medicine, 22,* pp. 125-133, 1992.
12. C. E. Land, Estimating Cancer Risks from Low Doses of Ionizing Radiation, *Science, 209,* pp. 1197-1203, 1980.
13. R. E. Shore, Occupational Radiation Studies: Status, Problems, and Prospects, *Health Physics, 59,* pp. 63-68, 1990.
14. E. S. Gilbert, Studies of Workers Exposed to Low Doses of External Radiation, *Occupational Medicine: State of the Art Reviews, 6,* pp. 665-680, 1991.
15. A. M. Stewart, Leukemia and Other Neoplasms in Childhood Following Radiation Exposure in Utero—A General Survey of Present Knowledge, *British Journal of Radiology, 41,* pp. 718-719, 1968.
16. A. Stewart and G. W. Kneale, Prenatal Radiation Exposure and Childhood Cancer, *Lancet, 1,* pp. 42-43, 1971.
17. T. F. Mancuso, A. Stewart, and G. Kneale, Radiation Exposures of Hanford Workers Dying from Cancer and Other Causes, *Health Physics, 33,* pp. 369-385, 1977.
18. V. Beral, P. Fraser, L. Carpenter, M. Booth, A. Brown, and G. Rose, Mortality of Employees of the Atomic Weapons Establishments, 1951-82, *British Medical Journal, 297,* pp. 757-770, 1988.
19. S. Wing, C. M. Shy, J. L. Wood, S. Wolf, D. L. Cragle, and E. L. Frome, Mortality among Workers at Oak Ridge National Laboratory: Evidence of Radiation Effects in

Follow-Up Through 1984, *Journal of the American Medical Association, 265,* pp. 1397-1402, 1991.

20. S. Wing, C. M. Shy, J. L. Wood, S. Wolf, D. L. Cragle, W. Tankersley, and E. L. Frome, Job Factors, Radiation and Cancer Mortality at Oak Ridge National Laboratory: Follow-up Through 1984 (published erratum appears in *American Journal of Industrial Medicine, 23*:4, p. 673), *American Journal of Industrial Medicine, 23,* pp. 265-279, 1993.
21. S. Wing, C. M. Shy, J. L. Wood, S. Wolf, D. L. Cragle, and E. L. Frome, Mortality among Workers at Oak Ridge National Laboratory, Evidence of Radiation Effects in Follow-up Through 1984, *Journal of the American Medical Association, 265,* pp. 1397-1402, 1991.
22. G. W. Kneale and A. M. Stewart, Factors Affecting Recognition of Cancer Risks of Nuclear Workers, *Occupational & Environmental Medicine, 52,* pp. 512-523, 1995.
23. H. Morgenstern, J. Froines, B. Ritz, and B. Young, *Final Report: Epidemiologic Study to Determine Possible Adverse Effects to Rocketdyne/Atomics International Workers from Exposure to Ionizing Radiation,* Contract No. 324A-8701-S0163, Public Health Institute, Berkeley, California, 1997.
24. D. B. Richardson and S. Wing, Methods for Investigating Age Differences in the Effects of Prolonged Exposures, *American Journal of Industrial Medicine, 33,* pp. 123-130, 1998.
25. D. B. Richardson and S. Wing, Greater Sensitivity to Radiation Exposures at Older Ages among Workers at Oak Ridge National Laboratory: Follow-up Through 1990, *International Journal of Epidemiology, 28,* pp. 428-436, 1999.
26. E. A. Dupree, J. P. Watkins, J. N. Ingle, P. W. Wallace, C. M. West, and W. G. Tankersley, Uranium Dust Exposure and Lung Cancer Risk in Four Uranium Processing Operations, *Epidemiology, 6,* pp. 370-375, 1995.
27. L. E. Sever, E. S. Gilbert, N. A. Hessol, and J. M. McIntyre, A Case-Control Study of Congenital Malformations and Occupational Exposure to Low-Level Ionizing Radiation, *American Journal of Epidemiology, 127,* pp. 226-242, 1988.
28. G. M. Kendall, C. R. Muirhead, B. H. MacGibbon, A. J. Conquest, A. A. Goodill, B. K. Butland, T. P. Fell, D. A. Jackson, M. A. Webb et al., Mortality and Occupational Exposure to Radiation: First Analysis of the National Registry for Radiation Workers, *British Medical Journal, 304,* pp. 220-225, 1992.
29. E. Cardis, E. S. Gilbert, L. Carpenter, G. Howe, I. Kato, B. K. Armstrong, V. Beral, G. Cowper, A. Douglas, J. Fix et al., Effects of Low Doses and Low Dose Rates of External Ionizing Radiation: Cancer Mortality among Nuclear Industry Workers in Three Countries, *Radiation Research, 142,* pp. 117-132, 1995.
30. E. S. Gilbert, D. L. Cragle, and L. D. Wiggs, Updated Analyses of Combined Mortality Data for Workers at the Hanford Site, Oak Ridge National Laboratory, and Rocky Flats Weapons Plant, *Radiation Research, 136,* pp. 408-421, 1993.
31. J. V. Neel, Problem of 'False Positive' Conclusions in Genetic Epidemiology: Lessons from the Leukemia Cluster near the Sellafield Nuclear Installation, *Genetic Epidemiology, 11,* pp. 213-233, 1994.
32. M. S. Lindee, *Suffering Made Real: American Science and the Survivors at Hiroshima,* University of Chicago Press, Chicago, Illinois, 1994.
33. J. W. Gofman, *Radiation-Induced Cancer from Low-Dose Exposure: An Independent Analysis,* Committee for Nuclear Responsibility, San Francisco, California, 1990.

34. E. L. Frome, D. L. Cragle, J. P. Watkins, S. Wing, C. M. Shy, W. G. Tankersley, and C. M. West, A Mortality Study of Employees of the Nuclear Industry in Oak Ridge, Tennessee, *Radiation Research, 148,* pp. 64-80, 1997.
35. R. S. Rabbitt, *Hotspots: The Legacy of Hiroshima and Nagasaki,* Cassell, London, 1995.
36. W. J. Schull, *Effects of Atomic Radiation: A Half Century of Studies,* Wiley-Liss, 1995.
37. R. Nussbaum and W. Köhnlein, Inconsistencies and Open Questions Regarding Low-Dose Health Effects of Ionizing Radiation, *Environmental Health Perspectives, 102,* pp. 656-667, 1994.
38. R. Sposto, D. O. Stram, and A. A. Awa, An Estimate of the Magnitude of Random Errors in the DS86 Dosimetry from Data on Chromosome Aberrations and Severe Epilation, *Radiation Research, 128,* pp. 157-169, 1991.
39. K. Neriishi, F. L. Wong, E. Nakashima, M. Otake, K. Kodama, and K. Choshi, Relationship Between Cataracts and Epilation in Atomic Bomb Survivors, *Radiation Research, 144,* pp. 107-113, 1995.
40. K. Neriishi, D. O. Stram, M. Vaeth, S. Mizuno, and S. Akiba, The Observed Relationship Between the Occurrence of Acute Radiation Effects and Leukemia Mortality among A-bomb Survivors, *Radiation Research, 125,* pp. 206-213, 1991.
41. S. Wing, C. M. West, J. L. Wood, and W. Tankersley, Recording of External Radiation Exposures at Oak Ridge National Laboratory: Implications for Epidemiological Studies, *Journal of Exposure Analysis and Environmental Epidemiology, 4,* pp. 83-93, 1994.
42. E. Cardis, E. Gilbert, L. Carpenter, G. Howe, I. Kato, J. Fix, L. Salmon, G. Cowper, B. Armstrong, V. Beral, A. Douglas, S. Fry, J. Kaldor, C. Lave, P. Smith, G. Voelz, and L. Wiggs, *Combined Analyses of Cancer Mortality Among Nuclear Industry Workers in Canada, the United Kingdom and the United States of America,* IARC Technical Report No. 25, World Health Organization, International Agency for Research on Cancer, Lyon, France, 1995.
43. G. W. Beebe, A Methodologic Assessment of Radiation Epidemiology Studies, *Health Physics, 46,* pp. 745-762, 1984.
44. A. M. Stewart and G. W. Kneale, A-Bomb Survivors: Further Evidence of Late Effects of Early Deaths, *Health Physics, 64,* pp. 467-472, 1993.
45. A. Stewart, Detection of Late Effects of Ionizing Radiation: Why Deaths of A-Bomb Survivors are So Misleading, *International Journal of Epidemiology, 14,* pp. 52-56, 1985.
46. A. M. Stewart and G. W. Kneale, Differences Between A-Bomb Survivors With and Without Acute Injuries, *European Journal of Oncology, 3,* pp. 361-366, 1998.
47. G. W. Beebe, C. E. Land, and H. Kato, The Hypothesis of Radiation Accelerated Aging and the Mortality of Japanese A-Bomb Victims, in *Late Biological Effects of Ionizing Radiation,* Vol. I, International Atomic Energy Agency, Vienna, pp. 3-27, 1978.
48. T. Wilcosky and S. Wing, The Healthy Worker Effect: Selection of Workers and Work Forces, *Scandinavian Journal of Work, Environment & Health, 13,* pp. 70-72, 1987.
49. G. S. Wilkinson, The SMR: Its Use and Misuse in Radiation Epidemiology, *CCRI June*, pp. 9-11, 1992.
50. H. M. Arrighi and I. Hertz-Picciotto, The Evolving Concept of the Healthy Worker Survivor Effect, *Epidemiology, 5,* pp. 189-196, 1994.

51. H. Checkoway, N. E. Pearce, and D. J. Crawford-Brown, *Research Methods in Occupational Epidemiology,* Oxford University Press, New York, 1989.
52. A. M. Stewart, J. Webb, D. Giles, and D. Hewitt, Malignant Diseases in Childhood and Diagnostic Irradiation in Utero, *Lancet, 2,* p. 447, 1956.
53. J. F. Bithell and A. M. Stewart, Pre-Natal Irradiation and Childhood Malignancy: A Review of British Data from the Oxford Survey, *British Journal of Cancer, 31,* pp. 271-287, 1975.
54. B. McMahon, Prenatal X-Ray Exposure and Childhood Cancer, *Journal of the National Cancer Institute, 28,* p. 1173, 1962.
55. E. B. Harvey, J. D. Boice, M. Honeyman, and J. T. Flannery, Prenatal X-Ray Exposure and Childhood Cancer in Twins, *New England Journal of Medicine, 312,* pp. 541-577, 1985.
56. T. Kuhn, *The Structure of Scientific Revolutions,* Vol. 2, University of Chicago Press, Chicago, 1962.
57. S. Aronowitz, *Science as Power: Discourse and Ideology in Modern Society,* University of Minnesota Press, Minneapolis, 1988.
58. H. J. Geiger, D. Rush, and D. Michaels, *Dead Reckoning: A Critical Review of the Department of Energy's Epidemiologic Research,* Physicians for Social Responsibility, Washington, D.C., 1992.
59. A. Makhijani, H. Hu, and K. Yih, *Nuclear Wastelands: A Global Guide to Nuclear Weapons Production and Its Health and Environmental Effects,* MIT Press, Cambridge, Massachusetts, 1995.
60. Advisory Committee on Human Radiation Experiments, *Final Report,* Advisory Committee on Human Radiation Experiments, U.S. Government Printing Office, Washington, D.C., 1995.
61. A. M. Stewart and G. W. Kneale, An Overview of the Hanford Controversy, *Occupational Medicine, 6,* pp. 641-663, 1991.
62. G. Wilkinson, Seven Years in Search of Alpha: The Best of Times, the Worst of Times, *Epidemiology, 10,* pp. 340-344, 1999.

Note: Originally published in *New Solutions, 9*(2), pp. 133-151, 1999.

CHAPTER 4

Science Is Not Sufficient: Irving J. Selikoff and the Asbestos Tragedy

Jock McCulloch and Geoffrey Tweedale

Professor Irving J. Selikoff (1915-1992) was America's foremost medical expert on asbestos-related diseases between the 1960s and early 1990s. He was also well known to the public for his media appearances on the burgeoning asbestos problem. Yet his reputation has been strikingly mixed. On the one hand, he has been portrayed as a mischief maker and irresponsible demagogue, who exaggerated the risks of asbestos and so destroyed an industry; on the other, as a pioneer in asbestos epidemiology, whose landmark studies of insulation (and other) workers demonstrated the severity of a modern occupational and public health tragedy. Drawing upon unprecedented access to the Selikoff archive at Mount Sinai Hospital in New York City, this chapter demonstrates that the most serious criticisms of Selikoff are either ill-founded or simply false. It also shows that Selikoff, in the highly politicized world of asbestos science, was a far more complex and conservative individual than previous studies have suggested.

> A major tragedy in American history is being played out before us. In a sense, we have all been on the stage—industry, government, legislators, Workers' Compensation Boards, industrial hygiene societies, insurers, courts, scientists. As act followed act, there have been recurring unrelieved scenes of inattention, unconcern, failure.
>
> — J. Selikoff to Senator Christopher Dodd, 23 October 1983, Hartford, Connecticut. Transcript titled, *The Asbestos Tragedy.*

Asbestos is the most physically damaging of the industrial minerals and most countries of the developed world have virtually banned its manufacture. It can cause incurable and often fatal diseases: notably asbestosis (lung scarring), lung cancer (originating in the lining of the airways of the lungs), mesothelioma (a virulent cancer of the lining of the chest or of the abdomen), and possibly other cancers. After more than a century of use, asbestos has left a legacy of litigation, bankrupted companies, and rising morbidity and mortality from asbestos-related disease (ARD) [1]. In the United States, an explosion in litigation has triggered demands for Congressional action. ARDs over the next decade or more will claim the lives of more than 100,000 Americans [2, 3]. No one knows the final global death toll, but the asbestos cancer epidemic may take upwards of five million lives before asbestos is banned world-wide [4].

These problems hit the news after the 1960s, when asbestos cancers—especially mesothelioma—were publicized. In America, a key figure in highlighting the asbestos hazard was a New York physician, Dr Irving Selikoff (1915-1992). Based at Mount Sinai Hospital, Selikoff emerged after the 1960s as America's most prominent and acclaimed researcher in ARDs [5]. He has had many admirers, not only among his medical peers, but also among activists and victims [6, 7]. But he also garnered controversy. In the 1960s, America was the world's largest asbestos user, with the leading asbestos company Johns-Manville listed as a *Fortune 500* company. The output of that industry—totaling thousands of products—penetrated into the heart of the American economy, especially into the construction and defense industries. Publicizing that asbestos was a mortal danger was fraught with difficulties for any medical researcher. Selikoff was soon portrayed as a media demagogue, who whipped up "fibre phobia" and asbestos "hysteria" on the back of bogus medical qualifications and flawed science. Since his death in 1992, the criticism has become even more vituperative and claims have persisted that he was a medical fraud [8-11].

However, none of these critical accounts have been remotely comprehensive. None have been based on a thorough review of the documentation that has become available since Selikoff's death through legal discovery. In particular, none have drawn on the Selikoff archive at Mount Sinai Hospital. In this chapter, we offer an assessment of Selikoff and his involvement with asbestos that for the first time scrutinizes Selikoff's own papers.

SELIKOFF AND ASBESTOS

Selikoff's work with asbestos first hit the headlines in 1964, when he appeared at an international asbestos conference in New York City (which he had taken a lead in organizing) [12]. Selikoff was then a relatively unknown 49-year-old New York physician. He was Brooklyn-born from an immigrant Jewish background and his early medical education was blighted by anti-Semitism and the Second World War, so that his medical diplomas had to be sought abroad. Licensed to practice in America, he became a hospital chest physician in New Jersey and New York and in the early 1950s was involved in the pioneering use of the tuberculosis drug isoniazid. Isoniazid brought Selikoff the first of his many honors—the Lasker Award (among whose recipients featured many future Nobel prize-winners)—and his first taste of the media limelight. A lucrative practice on Fifth Avenue might have beckoned, but instead in 1953 Selikoff founded a clinic in a New Jersey working-class community. It was there that he became aware of ARDs by observing workers from a local asbestos factory.

Selikoff wrote to the company asking for access to workers' medical records. When the company refused—and with no official help forthcoming either—in 1962 Selikoff contacted the local of the International Association of Heat & Frost Insulators & Asbestos Workers. Initially, they were suspicious, having been constantly reassured (and obstructed) by occupational health experts. However, Selikoff's tact and commitment won them over [6, 13]. With access to the insulators' work and medical profiles, Selikoff assembled a team for a scientific study. He was in an excellent position to do so, having become director of a new Environmental Sciences Laboratory at Mount Sinai Hospital in New York City. He recruited Dr. E. Cuyler Hammond (1912-1986), the American Cancer Society's director of statistics and epidemiology, who had already published a landmark study confirming the link between smoking and lung cancer [14]. Like Selikoff, he was interested in applying the latest epidemiological techniques to occupational cohorts—an idea that was still relatively new. Dr. Jacob Churg (b. 1910), a Mount Sinai pathologist, provided verification of the cause of death. What these individuals discovered was presented by Selikoff at the New York asbestos conference.

The conclusions from a dataset which involved a relatively small number of men (632) were conclusive. Insulation work was a death-trap. Selikoff's first study, published in 1964 and covering workers who were on the union rolls in 1943, showed that when these insulators were tracked to 1962 they had an excess death rate of 25%, with a heavier mortality than normal not only from asbestosis, but also lung cancer, mesothelioma, and stomach/colon/rectal cancer [15]. If this was alarming news for insulators, it was also damaging for the asbestos industry because it demonstrated how little interest the industry had shown in studying and protecting workers. Moreover, the independence and objectivity of Selikoff's studies meant that they could never be seriously challenged: even asbestos industrialists henceforth had to admit that insulation was a hazardous occupation.

In 1968, Selikoff and his team produced another classic study which illustrated the dangerous synergy between tobacco and asbestos, by showing that asbestos workers who smoked had 90 times the risk of developing an asbestos-related cancer than individuals who were non-smokers and were not exposed to asbestos [16]. A decade later, with the help of Herbert Seidman (another epidemiologist from the American Cancer Society), Selikoff and Hammond were able to provide further data on the smoking/asbestos risk, demonstrate that even short-term exposure to asbestos resulted in significant cancer risk up to thirty years later, and extend the cohort to include every man (17,800) in the insulators' union [17-19]. It was the largest study of asbestos workers worldwide, which Selikoff was to continue until his death. His team meanwhile undertook additional statistical studies, including projections by Dr. W. J. Nicholson on morbidity and mortality from ARD into the first decades of the 21st century [20]. The same team conducted important clinical field studies of asbestos-exposed workers in various trades, including the shipyards, Quebec asbestos mining, sheet-metal workers, and of household members of asbestos-exposed workers.

Perhaps Selikoff's longest-running battle concerned asbestos in the environment. At the start of the 1970s, Selikoff was drawn into disputes over the spraying of asbestos during the construction of the World Trade Center (WTC). This concluded in the banning of sprayed asbestos, first at the WTC, then throughout New York, and finally nationwide. During the 1980s, the environmental battle intensified over the issue of asbestos in office and school buildings. By 1986, President Reagan had signed an Abatement Act and the Asbestos Hazard Emergency Response Act (AHERA). It aimed to inspect schools for friable asbestos and introduce management plans for problem areas, with the intention to remove ("abate") asbestos if it was judged hazardous. The implications were enormous, not only for thousands of schools, but also for hundreds of thousands of commercial buildings throughout America.

Irving Selikoff, too, was worried about asbestos inside buildings, especially after 1986 when (at the request of the New York City Board of Education) he had examined school custodians and maintenance personnel and found that a significant proportion had asbestotic scarring at x-ray [21]. Selikoff viewed asbestos in buildings as potentially hazardous and believed that asbestos removal might sometimes be advisable. He was vigorously opposed not only by certain sections of the media, but also by property interests, the Canadian asbestos industry, and several scientists who believed that asbestos in buildings was innocuous. In 1990, Selikoff and his colleagues launched a major conference in New York, where participants argued that the risk of asbestos in buildings was real and that the exposure of schoolchildren, teachers, maintenance staff, and construction workers could result in a "Third Wave" of ARD [22]. It was the last conference that Selikoff attended and in which he had a major hand. Two further conferences he had planned on molecular biology took place without him. One of the last accounts of Selikoff depicts him screening asbestos workers day and night in 1991 at a trade union convention, even though he himself was sick [23].

SELIKOFF'S MODUS OPERANDI

The voluminous Selikoff archive, alongside interviews with close colleagues, form a picture of the man. His defining physical characteristic was energy: he could work 18-hour days and even during the final weeks of his life when hospitalized with pancreatic cancer he continued to write. To his immediate colleagues, he could sometimes be demanding and overbearing, often awakening them with dawn telephone calls. But he was never a careerist and though a dominant individual never used profane language [24]. Selikoff's public face was defined by the media: indeed, he has been described as the first media-savvy medical man. An asbestos public relations director paid tribute to Selikoff as a "Walter Cronkite of the scientific community," who almost invented the art of medical PR because he "always provided interesting, frequently controversial copy and he was always accessible . . . always patient, never condescending" [25].

Selikoff's media appearances meant that his motives were questioned; but his private face was much more complex. A colleague recalled that: "Like many Russian Jewish men, [Selikoff] had gargantuan visions, appetites and moods . . . [with] an enormous capacity for seeing the big picture" [26]. Selikoff saw himself as part of a rich medical tradition that went back to Bernardino Ramazzini and Sir William Osler. In the latter especially, Selikoff recognised a kindred spirit. He also greatly admired his American predecessors in occupational medicine such as Alice Hamilton, Wilhelm Hueper, and Harriet Hardy [27]. He felt the history of the discipline as a living presence and an inspiration.

Past masters influenced Selikoff in many ways. He often stated that "science is not sufficient"—a phrase that he appears to have drawn from Osler, and which recognized that occupational health was socially constructed. Selikoff saw that one of the perennial problems facing occupational medicine is the lag between the identification of hazards and reform of the workplace. Writing in 1986, he noted that over the previous two decades a great deal of data about exposure to toxic agents had been accumulated. Unfortunately, that information had not been fully utilized [28]. Selikoff partly blamed scientists like himself: "We (the profession) failed to protect [workers] and now disease and death are resulting. To add moral failure to the industrial failure, we are even reneging on the promises we made—that would allow the victims of our inadequacies at least the help they need and cannot earn themselves, and the dignity of dying with their family secure" [29]. He also recognized that part of the reason for that failure lay outside medicine. "I suspect that scientists have failed because they approach the problem of implementation of knowledge gained as a scientific matter. We did not realize that was rather only in part scientific, but at least in equal part a political and social problem" [30]. Of necessity, part of his role as a physician was educative and therefore political.

Selikoff also was much influenced by Osler's view that medical advances offered the prospect of prevention. As Selikoff observed: "Scientists like myself are 'keeping score,' producing good statistics in careful epidemiological studies, of people who are suffering and dying. [But] the brilliant progress that has been made in the past twenty years in molecular biology, cellular genetics and molecular biochemistry perhaps can give . . . an extraordinary opportunity, of which we are not taking advantage" [31]. To Selikoff, identification of high-risk groups (such as insulators) offered the possibility of treating cancers early by neutering the carcinogen. Thus, the "iron hand of latency" (another of his favorite phrases, which referred to the decades it took ARDs to develop), paradoxically offered hope. In 1989, he wrote ebulliently: "Our generation is the first to have the opportunity to prevent cancer! We can be optimistic. The problems are difficult, but the rewards will be great and the past two decades give reason to consider that we will be successful" [32]. Selikoff felt that despite the scale of the asbestos tragedy "individuals can make a difference" and so pessimism was unwarranted [33]. Optimism underpinned Selikoff's programs for mesothelioma treatment that were greeted so warmly by the unions because they offered hope for the sick and dying.

Selikoff became involved increasingly in medical research, but he retained a strong connection with working people and also was aware of the problems of the poor, the elderly, and black Americans. It was probably due partly to his New Jersey experiences; it also reflected the setting of Mount Sinai Hospital. What had begun as a fashionable urban hospital on Upper East Side Manhattan was by the 1960s flanked by poor working-class areas [34]. It was this changing social and physical environment that prompted Mount Sinai to form America's first urban department of community medicine, where Selikoff's Environmental Sciences Laboratory was located. Selikoff's persona within this milieu was that of the white-coated physician. Critics felt that it was a media pose, but Selikoff's skills and dedication were genuine. He has been described by one (by no means uncritical) admirer as a "consummate clinician and a great pulmonologist" [35]. He routinely went through x-rays in front of patients carefully explaining their meaning and the likely outcome. He was helped by a photographic memory for x-rays and case notes. He corresponded with the families of men and women suffering from asbestosis and mesothelioma. One file in the Selikoff archive is filled with harrowing letters—mostly written by grieving widows—that describe unspeakable suffering and deaths from ARDs. The letters invariably add thanks for Selikoff's help. He therefore understood directly the impact of occupational disease [27]. That distinguished him from asbestos researchers such as Sir Richard Doll and J.C. Wagner who never treated patients.

Empathy with the underdog underpinned Selikoff's rapport with the trade unions—a relationship that was occasionally tinged with reverence on the union side. This has led some to dub Selikoff a trade union crusader, who was bankrolled by labor. But Selikoff maintained a professional distance from the unions. He commented once that his group's support came from a wide variety of national institutes and official bodies and added: "Our research studies are independent, are not done "for" government or trade unions or individuals. . . . Our doors are open, however, to those who can benefit from our knowledge, including trade unions, government bodies, individuals, industrial groups, etc." [36]. Selikoff masterminded programs—such as the Insulation Industry Hygiene Research Program in 1968—that initially drew on union support, but Selikoff himself never received any funds to direct or work with the program. And he was always careful to involve industry (in this case, Johns-Manville) and government, besides the unions. Within months, Selikoff indicated that it was unfair for the ordinary worker to support the program with further contributions [37]. Selikoff stated [36]:

> We do not receive money from trade unions, although we do not think that this is a bad idea for others, in some circumstances. . . . [However] we are aware of trade union funds made available "without strings" with dubious results and, in other circumstances, arrangements being negotiated by labor and industry and jointly approved, with sometimes uncertain success. It may well be that trade union groups have not yet always developed the sophistication needed for evaluation of proposals. Indeed, consultation with academic units can be an important service to such trade unions (and to industry, often, as well).

Selikoff preferred to work with all interested parties. Often, it was at their request. One example is his unit's survey of white asbestos (chrysotile) miners in Baie Verte, Newfoundland, which was completed in 1978. The initial approach was made by the provincial government, but Selikoff soon had on board Johns-Manville, the leading labor union (the United Steelworkers), and the Canadian government. Selikoff, though, retained the right to publish as he saw fit. At the end of the survey, Selikoff sent a copy to Johns-Manville medical director Paul Kotin, thanking the company for its assistance [38]. Despite his links with the unions, Selikoff's attitude to industry was always cordial and his strategy was to educate management. He believed that when faced with scientific evidence of the hazards of asbestos the men who ran global corporations would improve the conditions in their factories [24]. Selikoff expected a modicum of good faith on the other side. That was not always forthcoming and companies like Johns-Manville simultaneously worked with him and conspired to undermine him.

In 1968, he stated that it was "idle and largely inaccurate to 'blame the industry' for the present situation"—and rejected what he described as the "demonological theory" of industrial hygiene history [39]. That was the year when the Insulation Industry Hygiene Research Program was launched by Mount Sinai Hospital in association with Johns-Manville. In an article publicizing the partnership, Selikoff stated: "It is much nearer the truth to say that both industry and labor shared in ignorance and neglect of the problem. Science and medicine are also at fault here for inadequate attention to environmental and occupational health" [40]. Among trade union friends, Selikoff's comments were no less conciliatory toward industry. After surveying the disturbing numbers of industrial deaths in the US, he told one gathering: "Remember the story of who killed Cock Robin? Well . . . No one killed Cock Robin, no one. Inattention killed Cock Robin, research that wasn't done, meetings that weren't held, training that was never given, laws that were never passed. Omission, unconcern killed Cock Robin" [41].

Sometimes Selikoff's views were close to those of the asbestos industry. He recognized that asbestos had been a useful material and, initially, he distrusted substitute materials such as fiber glass observing that, "There is insecurity about much of the data available so far concerning their potential for the induction of cancer" [42]. He also favored controlling asbestos hazards, rather than campaigning for an outright ban. This stance was due to Selikoff's acceptance of political realities—in America hazardous materials were historically controlled rather than banned—but also reflected his belief that control measures should be given a chance. He told the British press in 1977 that he was not in favor of a ban [43]. In the following year, Selikoff and Mount Sinai scientists published a major textbook, *Asbestos and Disease* [44]. It began with a foreword by Paul Kotin, contained no calls for banning asbestos, but did have sections on dust control illustrated with (not entirely convincing) photos of workers mixing asbestos-cement in enclosed bags and spraying asbestos within sealed areas. The book ended with a section entitled "Cooperative Effort," which had an upbeat view of the relationship between industry, labor, and government.

In contrast to the industry, which refused to take part in transparent public debate, Selikoff always sought to engage with the science of his opponents. In November 1985, he wrote to Dr. Jacques Dunnigan (a long-term supporter of Canadian chrysotile) chiding him for not attending a conference at Mount Sinai. "While I understand that an important goal of [your] research is to find possible ways to more safely use asbestos in the future, I believe it equally urgent to study how to ameliorate the disastrous consequences of past inadequacies. It is for these reasons that we so very much hoped you would accept an invitation. And why we so very much regretted that neither you nor your colleagues came" [45]. Selikoff went on to comment that at a recent scientific meeting European colleagues had damned Canadian scientists as "salesmen for asbestos." "I felt that was irrelevant, and still do," wrote Selikoff: "I considered that if research would lead to better understanding of the problem, that would be sufficient. For this reason, I hope that in your own way and at your own time, you will use your research to assist in disease control. I am sure you shall" [45].

Litigation never disturbed Selikoff's attempts to get the industry onside. He was pained at how the statutory workmen's compensation system in America had failed asbestos victims. Equally, he was aware of the swelling tide of court cases. One might have expected him to become a leading expert witness for plaintiffs—indeed one study [10] has suggested that he did—yet in the legal sphere Selikoff avoided the limelight. Selikoff gave evidence in some of the first landmark cases in the late 1960s—such as the Claude Tomplait case [7, 46]—but by the early 1970s Selikoff had decided to shun litigation. He recognized that it would be an enormous drain on his time and would push him into confrontation with the industry. Selikoff also had custody of medical and union records, which—to retain the trust of labor—had to remain strictly confidential. Requests for legal help invariably received a polite refusal. He replied to one such request in

1983: "because of teaching and research obligations, I have kept the firm policy of not participating in medico-legal matters for many years. Recognizing this consistency, the courts have respected my constraints" [47]. This position did not prevent Selikoff from avidly reading courtroom transcripts and pondering deeply the social problems of compensation and litigation. His reaction was characteristic: he organized conferences on compensation; encouraged attorneys of all shades to attend special courses at Mount Sinai; and remained on friendly terms with leading attorneys and judges. Selikoff was detached enough to recognize both the merits of litigation in winning compensation and also its deficiencies. Typically, he was once attracted to the idea of a special fund for asbestos victims that would bypass litigation, even writing to Warren Burger—a conservative judge—to congratulate him on his plans for an alternative to litigation [48].

Selikoff's dissemination of scientific information operated on a global scale. Using a grant from the Environmental Cancer Information Centre, and employing a team of five secretaries to type his voluminous correspondence, he became the center of a global network of occupational health scientists, public health officers, and trade unionists. Selikoff circulated the latest data to colleagues around the globe thereby enabling colleagues, especially in developing countries, to keep abreast of the science. Myra Karstadt for a time coordinated that work at Mount Sinai [24]. In an era before electronic data bases, it was a significant innovation.

The Collegium Ramazzini was established by Selikoff and a small group of colleagues in 1982. The Collegium was a conscious attempt to bring occupational health issues into a global context covering both developed and developing countries [27]. It grew out of research on vinyl chloride conducted by Selikoff and the Italian scientist Cesare Maltoni, which led them to organize a conference on vinyl chloride in New York in 1975. They discussed the need for an independent organization which could duplicate the global structure of corporations, but was aimed at combating the spread of industrial hazards and fostering the acquisition of medical knowledge.

The Collegium enabled Selikoff to disseminate knowledge world-wide. In 1987, for example, Selikoff wrote to Mikhael Gorbachev [49]. In the 1980s, the USSR was the world's largest chrysotile user and also exported large quantities of fiber. Selikoff pointed out the international concern among scientists about asbestos. In technologically advanced economies like the USSR and the United States, the means existed to control exposure. "This is not the case in the developing countries," wrote Selikoff, "where the infrastructure needed for hazard control is embryonic, fragmentary or, in some cases, almost non-existent. It has been depicted as "exporting cancer," although the resulting disease will not be seen for 20, 30, or more years" [49]. Selikoff hoped that Gorbachev would find time to review this matter with Soviet experts and offered the assistance of the Collegium.

SAINT OR SINNER?

Some industrialists warmed to Selikoff. One asbestos plant manager attended a three-hour talk that Selikoff presented to the Asbestos Information Association in New York in 1974 and commented: "I was very impressed by him and his sincerity and dedication to his studies. I had a misconception of his motives and activities before this meeting. I talked to him, personally, during lunch and I believe his motives are aboveboard . . ." [50]. Yet many industrialists and scientists distrusted Selikoff and tried to discredit him. The media were fed stories that Selikoff's medical qualifications were bogus: a smear that is still being recycled by asbestos industry consultants [10]. Given Selikoff's conservative approach and conciliatory attitude, that in turn needs explanation.

Partly, the animosity reflected Selikoff's dominance of the scientific discourse on asbestos, which overshadowed the public relations efforts of the industry. The industry regarded him as a "colossus," bestriding the asbestos scene and causing a "curious air of despondency" among American companies [1]. The industry never forgave Selikoff for publicizing the plight of insulation workers. Moreover, although he was sympathetic to the predicament of the industry, Selikoff was unflinching in describing the suffering of those workers and frequently drew attention to it. For example, he began one talk by telling his audience, "I don't think I'm going to be the most popular speaker on your program. I don't think I will exude a good deal of confidence. I am just not in the mood today." The reason was that Selikoff had just visited the hospital to see John Evilege, an insulator dying of lung cancer. Evilege's fate became the keynote of Selikoff's speech [41].

Selikoff continually trod on the industry's toes, whether he was investigating the hazards of sprayed asbestos, flagging the dangers of women's coats that contained asbestos fiber, picking holes in UK government safety thresholds, or raising the issue of asbestos in drinking water or in brake linings. Above all, Selikoff lifted the lid of the Pandora's Box of asbestos in the environment, where the industry's greatest liabilities lay. Moreover, although Selikoff favored controlling rather than banning asbestos, his support for such a policy was qualified. He warned the industry that "where control measures were available, but were not applied, the question of discontinuance of such use would have to be considered, and where control measures were not available and where significant emissions could be expected, such use would, once more, have to be looked at critically" [51].

Selikoff would have no truck with some of the industry's most cherished beliefs, particularly the view that chrysotile was harmless, when compared to crocidolite (blue) and amosite (brown) asbestos [52]. This issue became a constant source of friction between Selikoff and the industry, particularly in Canada, where industrial interests supported the work of Dr. Corbett McDonald at McGill University. Selikoff refused to condemn McDonald for building a career on industry sponsorship. "I have found him very knowledgeable," Selikoff told one correspondent, "and I do not believe that what he says can be disregarded merely because of his industry associations. Rather, his statements should be examined as they stand and evaluated as serious scientific questions" [53]. Despite this stance, relations between Mount Sinai and McGill cooled during the 1980s and by 1990 hostilities had been declared. After Selikoff's death, the McGill scientists attacked Selikoff's group for their "menace" and "intense *malice*" [8].

For his part, Selikoff became increasingly sensitized to industry corruption and malpractice and his correspondence contains occasional asides on the subject. He complained in 1973 of the heavy concentration of asbestos industry personnel and consultants in the composition of an International Agency for Research on Cancer meeting in Lyon and especially at the "extraordinary selection of an industry representative to give the only address at the conference dinner (paid for by the asbestos industry!), in which he regretted the fact that some must die in the industrial utilization of asbestos. But, since society needed this valuable material, we had to be brave and accept this unfortunate necessity" [54]. Selikoff disassociated himself from the proceedings. He was aware that many of his contemporaries were compromised by their relations with industry (though he would not have known that even major figures such as Sir Richard Doll and J.C. Wagner had secret and lucrative consultancies with the companies, whose products they then defended) [55, 56]. In reference to Dr. Robert Murray, a UK factory inspector who subsequently appeared as an expert witness for the asbestos industry, Selikoff wrote: "The actions of a relatively few individuals are serving . . . to cheapen a tradition that so many of us had held valuable . . ." [57]. Selikoff noted the predominance of industry consultants when the Health Effects Institute was formed in 1990 to assess the dangers of asbestos in buildings [58]. He deplored the "tendency nowadays for research to follow and be organized by agencies established with private support (whether they be "Institutes" in Boston or "Foundations" in Miami or whatever). . . . The scale, the scope of this new development in science is not derived from the philosophy of Academe" [59].

But Selikoff's touch was not infallible, especially in his dealings with industry scientists. Paul Kotin had joined Johns-Manville in the early 1970s, where he headed up its response to the asbestos crisis. Kotin had held senior positions as a pathologist with the University of Southern California and with the National Cancer Institute. Selikoff respected him as a scientist and he wrote of Kotin in 1978: "Paul is in the unusual position of being a very good scientist and a company medical director simultaneously. I would much rather that he be at Johns-Manville, than that he not be. In his position there, he has apparently found it necessary to guide the company in their opposition to workers' claims, and to testify for them accordingly. I've not read his testimony, and deal with him on a scientific level, rather than as the company doctor" [60]. It was an unlikely friendship, given Kotin's track record in the industry and his work as a tobacco industry consultant [61].

Associating with industry physicians sometimes caused outright misjudgements. In December 1984, Selikoff learned of the death of Dr. Walter Smither, who for many years was the senior medical officer for the British company Cape Asbestos. Selikoff wrote to a mutual friend, Dr. Molly Newhouse: "It was an unhappy moment to read of Walter Smither's death. I'm not sure that his contribution to the public's health is as well-known as it perhaps should be. Would you write a Commemoration for the *American Journal of Industrial Medicine*? He was a good model for industry physicians" [62]. Smither was in fact a sinister figure whose medical career was strewn with corpses. For more than two decades, Smither participated in Cape's appalling treatment of its black and colored workers in South Africa. Unbeknownst to his colleagues in the United States, Smither suppressed vital evidence of mesothelioma among Cape's employees and made no protest to the company's board that Cape workers were dying of scurvy [63].

Inevitably, Selikoff's involvement with the media, while it educated the public about occupational disease, also carried risks. Statements could be taken out of context and he had little control over how they were edited. As a result, he was often criticized for his public profile and particularly for the way in which he used the press, sometimes theatrically. In 1978 Selikoff gave several interviews during which he emphasized the dangers of working with asbestos. Harriet Hardy wrote to him in protest: "I refer to my reaction to the three or four national network TV viewings of the predicament of asbestos workers. I find it incredible that a man of your sensitivity, well-trained in medicine would not be more careful of the sorrow and panic that will come to the viewers of these programs, thinking of themselves and their families" [64]. Selikoff admired Hardy and her letter must have caused him some discomfort. He replied, not altogether convincingly: "I'm afraid you give me credit for omniscience that I don't have. I'm not infrequently asked to comment. Precisely because of the hope of having as much accuracy as possible I accede. Commonly, much of what I say is left on the cutting room floor" [65].

EVALUATION

Selikoff's promethean career is not easily assessed in short compass, but his "voice"—with its mix of science, ethics, concern for the sick and poor, and unsparing analysis of the asbestos tragedy—may be unique in occupational health. His purely scientific contributions are substantial enough to silence any but the most prejudiced of his critics; and our appraisal of his papers demonstrates that the most serious criticisms of Selikoff are either ill-founded or simply false.

Isoniazid remains a mainstay of tuberculosis treatment in combination with other drugs. The work of Selikoff and his collaborators on U.S. insulators had been accepted as a classic contribution to epidemiology. First published in the *Journal of American Medical Association* in 1964, this work was reproduced as a landmark article in *JAMA* in 1984 and was also reprinted as a "classic in oncology" in *CA* in the same year [66, 67]. The work of the same team in demonstrating the synergy of asbestos with smoking also has become a fundamental part of the asbestos literature. These landmarks stand alongside the studies of Merewether on asbestosis (1930), Doll on asbestos-related lung cancer (1955), and Wagner on mesothelioma (1960). In the wider history of epidemiology, the research inspired by Selikoff has been ranked alongside the work of Robert Case on occupational bladder cancer and Sir Richard Doll on smoking and lung cancer [68, 69]. The statistical estimates of future ARD mortality that Selikoff's unit generated in 1982 (courtesy especially of Bill Nicholson) have also stood up well: when the RAND Institute for Civil Justice conducted a sample comparison of the Nicholson projections with the best available recent data on U.S. mesothelioma it was found that the two data sets matched quite well. RAND noted that litigators still used the Nicholson study as "a standard reference on occupational exposure to asbestos" [3].

Certainly, some of Selikoff's anxieties and hypotheses have not transpired into major problems (something that would have given him pleasure rather than disappointment); others have had mixed fortunes in the light of recent research. Selikoff's identification of a linkage between asbestos and gastrointestinal cancer now looks less sure, with several studies (without completely rejecting the Selikoff hypothesis) unable to establish a causal link [70]. Sufficient evidence of a health hazard caused by asbestos in water has not been found—or is too small to be quantifiable [71]. Selikoff's distrust of man-made mineral fibers (MMMFs) now looks too guarded in the light of the WHO's recommendation that they should be used, where possible, as a substitute for asbestos (though the consensus of WHO discussions is that MMMFs cannot be deemed totally innocuous). Selikoff's view that chrysotile was an essentially similar health hazard to other commercial asbestos fibers has not been universally accepted, with many scientists arguing that there is a gradient of risk, with crocidolite and then amosite being more toxic than chrysotile. On the other hand, by the early 21st century most governments in the developed world, especially in Europe (though not Canada), had banned all forms of asbestos, including chrysotile. The WHO (and other global organizations) have recently underlined the carcinogenic risks of chrysotile and stated that no safe threshold has been identified [72-74].

Selikoff's emphasis on the environmental impact of asbestos and the emergence of the Third Wave now looks prescient. The highest proportion of deaths from mesothelioma is now among construction workers and tradesmen, such as carpenters and plumbers. The denigration of Selikoff's warnings of the potential hazards of asbestos in buildings and in schools has not made the problem disappear. In 2004, an American study highlighted that "elementary schoolteacher" was the eighth most frequently cited occupation on mesothelioma death certificates [2]. In 2006, the UK government introduced measures to reduce the risk from asbestos in buildings; it also warned of the dangers of classroom asbestos, because statistics showed that a disproportionate number of schoolteachers had died of mesothelioma [75, 76].

In the wider scheme of occupational health, the Collegium Ramazzini is still active, alongside the medical journals (such as the *American Journal of Industrial Medicine*) that Selikoff helped found. Today the Collegium has members in 40 countries and addresses the risks of disease and injury from the workplace and the environment. The Collegium has been successful in issuing authoritative statements on hazards. Its most effective interventions have been in calling for an international ban on asbestos and supporting countries such as France in resisting Canadian challenges under the WTO appeal process [77]. The Mount Sinai Environmental Health Sciences Center remains a key part of the New York medical scene. Among its most recent work has been assessing the impact of the toxic dusts released during the World Trade Center collapse [78]. The Center also has launched many careers in occupational medicine, with Selikoff's protégés scattered around the United States and the world [79]. Above all, Selikoff's emphasis on the environment and his focus on the problems of "living in a chemical world" have ensured that his work has proved highly relevant for a present era beset by environmental pollution, global warming, and the problems of the developing world.

Our respect for Selikoff has increased as we have studied his career, while our opinion of his critics has diminished. He was not infallible and he could be exasperating to his colleagues: though without his personality and media skills he would not have gotten his message across. The picture that emerges of Selikoff—from his personal papers, publications, and reminiscences of colleagues—is consistent and totally opposite to that projected by the asbestos industry and its sympathizers. He always put achieving his key goals—work safety and reducing disease—ahead of any personal sense of distaste. He was able to recognize the achievements of others and he seems to have been free of the petty malice that is so common in Academe. His correspondence reveals a man who spoke to everyone in the same measured voice, with no profanity or bitterness. Overall, he was a pragmatist and not the man who destroyed an industry. His preferred method was to work from the inside to bring about change within a system, which (as he well knew) was marked by a grossly unequal power relationship between American industry and labor. Selikoff did not politicize the science of ARDs (though this was a criticism constantly leveled at him): that had already been accomplished by the industry from the 1930s.

Ironically, Selikoff may have put too much faith in research. His hope that asbestos in the lung could somehow be inactivated by other agents; or that the identification of high-risk individuals could help reduce mortality have both proved chimerical. Today the worst asbestos cancers remain as resistant to treatment as in Selikoff's day. On his deathbed, apparently he regretted that he had waited so many years to get asbestos banned and wondered whether the years he had spent in research were misspent: "I could have saved a lot of those boys instead of waiting" [80]. Provocative as it may sound, when discussing a man who has emerged as a hero for the ban-asbestos movement, it could be that, far from destroying the industry, Selikoff may have been too conservative in his methods.

ACKNOWLEDGMENTS

The authors would like to thank Professor Phillip Landrigan, Professor Stephen Levin, and Ms. Valerie Josephson of Mount Sinai Hospital for their assistance.

REFERENCES

Documents cited in this chapter are from the Selikoff archive, Mount Sinai Hospital, New York City, unless otherwise stated.

1. B. I. Castleman, *Asbestos: Medical and Legal Aspects*, New York, Aspen Publishers, 5th edition, 2005.
2. Environmental Working Group, *Asbestos: Think Again*, Washington, D.C., 2004. Posted at: www.ewg.org/reports/asbestos
3. S. J. Carroll, D. Hensler, J. Gross, E. M. Sloss, M. Schonlau, A. Abrahamse, and J. S. Ashwood, *Asbestos Litigation,* Santa Monica, CA: RAND, pp. 16-18, 2005.
4. J. LaDou, The Asbestos Cancer Epidemic, *Environmental Health Perspectives*, *112*, pp. 285-290, 2003.
5. P. J. Landrigan, In Memoriam: Irving J. Selikoff, 1915-1992, *American Journal of Public Health*, *83*, p. 609, April 1993.
6. P. Brodeur, *Expendable Americans*, New York, Viking, 1974.
7. P. Brodeur, *Outrageous Misconduct*, New York, Pantheon Books, 1985.
8. F. D. K. Liddell, Magic, Menace, Myth and Malice, *Annals Occupational Hygiene*, *41*, pp. 1-12, 1997.
9. J. K. Corn, *Environmental Public Health Policy for Asbestos in Schools: Unintended Consequences*, Boca Raton, Lewis Publishers, 1999.
10. P. Bartrip. Irving John Selikoff and the Strange Case of the Missing Medical Degrees, *Journal of History of Medicine and Allied Sciences*, *58*, pp. 3-33, 2003.
11. D. Egilman, G. Tweedale, J. McCulloch, W. Kovarik, B. I. Castleman, W. Longo, S. Levin, and S. R. Bohme, P. J. W. Bartrip's Attack on Irving J. Selikoff, *American Journal of Industrial Medicine*, *46*, pp. 151-155, 2004.
12. M. Greenberg, Biological Effects of Asbestos: New York Academy of Sciences 1964, *American Journal of Industrial Medicine 43*, pp. 543-552, 2003.
13. J. Churg, Interview by Roland Blantz, University of California at San Diego, La Jolla, California. ISN Video Legacy Project. 22 May 1977. Posted at: http://www.cybernephrology.ualberta.ca/ISN/VLP/Trans/Churg.htm
14. L. Garfinkel, Asbestos: Historical Perspective, *CA: A Cancer Journal for Clinicians*, *34*, pp. 44-47, 1984.
15. I. J. Selikoff, J. Churg, and E. C. Hammond, Asbestos Exposure and Neoplasia, *Journal of American Medical Association, 188,* pp. 22-26, 1964.
16. I. J. Selikoff, E. C. Hammond, and J. Churg, Asbestos Exposure, Smoking, and Neoplasia, *Journal of American Medical Association*, *204*, pp. 106-112, 1968.
17. E. C. Hammond, I. J. Selikoff, and H. Seidman, Asbestos Exposure, Cigarette Smoking, and Death Rates, *Annals New York Academy of Sciences*, *330*, pp. 473-90, 1979.

18. H. Seidman, I. J. Selikoff, and E. C. Hammond, Short-Term Asbestos Work Exposure and Long-Term Observation, *Annals New York Academy of Sciences, 330*, pp. 61-89, 1979.
19. I. J. Selikoff, E. C. Hammond, and H. Seidman, Mortality Experience of Insulation Workers in the United States and Canada, 1943-1976, *Annals New York Academy of Sciences*, *330*, pp. 91-116, 1979.
20. W. J. Nicholson, G. Perkel, and I. J. Selikoff, Occupational Exposure to Asbestos: Population at Risk and Projected Mortality 1980-2030, *American Journal of Industrial Medicine*, *3*, pp. 259-311, 1982.
21. I. J. Selikoff, letter to Laura Job, NYS Department of Labor, Brooklyn, New York, 31 March 1988.
22. P. J. Landrigan and H. Kazemi (eds.), The Third Wave of Asbestos Disease: Exposure to Asbestos in Place. Public Health Control, *Annals of New York Academy of* Sciences, *643*, pp. 1-628, 1991.
23. S. W. Samuels and A. C. Upton (eds.), *Genes, Cancer, and Ethics in the Work Environment*, Solomons Island, Maryland, Ramazzini Institute/OEM Press, 1998.
24. Philip Landrigan, interviewed by J. McCulloch, Mount Sinai Hospital, 17 November 2006.
25. M. M. Swetonic, Death of the Asbestos Industry in *Crisis Response; Inside Stories of Managing Image under Siege*, J. Gottschalk (ed.), Detroit, Gail Research Inc., pp. 289-308, 1993.
26. E. D. Richter, personal communication to G. Tweedale, 17 December 2006.
27. Stephen Levin, interviewed by J. McCulloch, Mount Sinai Hospital, 27 October 2006.
28. I. J. Selikoff, letter to L. L. Moerwald, 11 August 1986.
29. I. J. Selikoff, The Asbestos Tragedy. Statement to Senator Christopher Dodd, Hartford, Connecticut, 23 October 1983.
30. I. J. Selikoff, Statement to Subcommittee on Health & Safety of the Committee on Education and Labor (chairman Joseph M. Gaydos), Washington, 17 July 1985.
31. I. J. Selikoff, letter to Senator Lowell Weicker, 2 May 1989.
32. I. J. Selikoff, letter to Professor Keith Griffiths, Tenovus Institute for Cancer, 6 April 1989.
33. I. J. Selikoff, letter to Professor Dan Mintz, 18 October 1991.
34. Anon, Environmental Science in the City that Never Sleeps, *Environmental Health Perspectives 103*, pp. 654-659, 1995.
35. Arthur Frank, interviewed by G. Tweedale, 18 December 2006.
36. I. J. Selikoff, letter to Peter Visser, 5 April 1982.
37. Anon, letter headed "Dear Sirs and Brothers", c. 1970.
38. I. J. Selikoff, letter to P. Kotin, 24 January 1978.
39. I. J. Selikoff, Statement, 7 March 1968, before the House Committee on Education and Labor, Select Sub-Committee on Labor, HR 14816, Occupational Safety & Health Act of 1968.
40. I. J. Selikoff, Partnership for Prevention—The Insulation Industry Hygiene Research Program, *Industrial Medicine*, *39*, pp. 21-25, 1970.
41. I. J. Selikoff, Address to the Second International Safety Conference, United Steel Workers of America, AFL-CI0. Conrad Hilton Hotel, Chicago. March 25, 26, 27, 1971.
42. I. J. Selikoff, letter to M. Toljagic, 13 May 1991.
43. I. J. Selikoff, interview with *Sheffield Star*, December 1977.

44. I. J. Selikoff and D. H. K. Lee, *Asbestos and Disease*, New York, Academic Press, 1978.
45. I. J. Selikoff, letter to Dunnigan, 7 November 1985.
46. I. J. Selikoff, Deposition in *Claude J. Tomplait v. Combustion Engineering,* US District Court for Eastern District of Texas Beaumont Division, Civ. No. 5402, New York, 4 March 1968.
47. I. J. Selikoff, letter to Alan Dalton, 24 October 1983.
48. I. J. Selikoff, letter to Chief Justice Warren Burger, 5 October 1982.
49. I. J. Selikoff, letter to Gorbachev, General Secretary, Kremlin, Moscow, 22 June 1987.
50. Horace Beasley, memo 25 February 1974 re. AIA/NA meeting 21 February 1974 to Capco (Birmingham, Alabama). Copy provided to authors by David Egilman.
51. I. J. Selikoff, letter to William P. Raines, AIA/NA, 2 April 1971.
52. G. Tweedale and J. McCulloch, Chrysophiles versus Chrysophobes: The White Asbestos Controversy, 1950s-2004, *Isis*, *95*, pp. 239-59, 2004.
53. I. J. Selikoff, letter to P. K. Kelly, 13 February 1978.
54. I. J. Selikoff, letter to Sheldon Samuels, AFL-CIO, 29 May 1973.
55. L. Hardell, M. J. Walker, B. Walhjalt, L. S. Firedman, and E. D. Richter, Secret Ties to Industry and Conflicting Interests in Cancer Research, *American Journal of Industrial Medicine*, *50*, pp. 227-233 2007.
56. J. McCulloch, Saving the Asbestos Industry, 1960 to 2003, *Public Health Reports*, *121*, pp. 609-614, 2006.
57. I. J. Selikoff, letter to M. Greenberg, 31 December 1979.
58. I. J. Selikoff, letter to Dr Arthur C. Upton, 14 May 1990.
59. I. J. Selikoff, letter to Arthur C. Upton, 20 December 1989.
60. I. J. Selikoff, letter to Greenberg, 28 February 1978.
61. Richard Kluger, *Ashes to Ashes: America's Hundred-Year Cigarette War, the Public Health, and the Unabashed Triumph of Philip Morris*, New York, Alfred A. Knopf, pp. 207, 325, 361, 1996.
62. I. J. Selikoff, letter to Dr. M. Newhouse, 12 December 1984.
63. J. McCulloch, *Asbestos Blues: Labour, Capital, Physicians and the State in South Africa*, London, James Currey, 2002.
64. H. Hardy, letter to Selikoff, 26 July 1978.
65. I. J. Selikoff, letter to Hardy, 2 August 1978.
66. I. J. Selikoff, J. Churg, and E. C. Hammond, Asbestos Exposure and Neoplasia, *Journal of American Medical Association*, *252*, I, 41-51, 1984. Reprinted from *Journal of American Medical Association, 188*, pp. 22-6, 1964.
67. I. J. Selikoff, J. Churg, and E. C. Hammond, Asbestos Exposure and Neoplasia. *CA: Cancer Journal for Clinicians*, *34*, pp. 48-56, 1984. Reprinted from *Journal of American Medical Association*, *188*, pp. 22-26, 1964.
68. S. D. Stellman, Issues of Causality in the History of Occupational Epidemiology, *Sozial-und Praventivmedizin, 48,* pp. 151-161, 2003.
69. M. Gochfield, Chronological History of Occupational Medicine, *Journal of Occupational and Environmental Medicine, 47,* pp. 96-114, 2005.
70. R. W. Morgan, D. E. Foliart, and O. Wong, Asbestos and Gastrointestinal Cancer, *Western Journal of Medicine*, *143*, pp. 60-65, 1985.
71. World Health Organization, *Asbestos in Drinking Water, WHO/SDE/WSH/03.04/02*, Geneva, WHO, 2003.

72. International Programme on Chemical Safety, *Environmental Criteria 203: Chrysotile Asbestos*. Geneva, WHO, 1998. Posted at: www.inchem.org/documents/ehc/ehc/ehc203.htm
73. World Trade Organization, *European Community—Measures Affecting Asbestos and Asbestos-Containing Products, Report of the Panel, WT/DS135®*, 18 September 2000.
74. World Trade Organization, *European Community—Measures Affecting Asbestos and Asbestos-Containing Products. Report of the Appellate Body, WT/DS135/AB®*, 12 March 2001.
75. Health and Safety Executive, *Asbestos—Updated Information for Schools*, 3 April 2006. Posted at: http://www.hse.gov.uk/press/2006/e06040.htm
76. M. Lees, *Asbestos in Schools*, private publication, 2006. Posted at: http://www.asbestosexposureschools.co.uk
77. B. I. Castleman, WTO Confidential, *International Journal of Health Services*, *32*, pp. 489-501, 2002.
78. P. J. Landrigan, et al., Health and Environmental Consequences of the World Trade Center Disaster, *Environmental Health Perspectives*, *112*, pp. 731-739, 2004.
79. R. Lilis, Award Acceptance Speech, *American Journal of Industrial Medicine, 49*, pp. 705-707, 2006.
80. E. D. Richter and R. Laster, The Precautionary Principle, Epidemiology and the Ethics of Delay, *Human and Ecological Risk Assessment*, *11*, pp. 17-27, 2005.

Note: Originally published in *New Solutions, 17*(4), pp. 293-310, 2007.

CHAPTER 5

Silenced Science: Air Pollution Decision-Making at the EPA Threatens Public Health

Kathleen Rest

The saga of the Environmental Protection Agency's new particulate matter (PM) rule is yet another example of this Administration's disregard for and disrespect of science and scientists—and may signal the beginning of a disturbing trend to reduce the role of science in protecting the quality of our air. Political interference in the PM case is clear. And more trouble may be in the wings when it comes to acceptable levels of ozone pollution and the process for setting the National Ambient Air Quality Standards (NAAQS). For several years, the Union of Concerned Scientists has been actively monitoring and documenting the misuse of science in public policy-making. Consider this a call to arms. Now is the time to engage your elected officials on these issues.

When reading about the Environmental Protection Agency's (EPA) recently finalized rule on particulate matter (PM) pollution chronicled on the Union of Concerned Scientists (UCS) web site and reprinted in this issue, one might view this as an unfortunate but isolated incident of politics trumping science in the policy-making process. Unfortunately, the saga of the new rule is yet another example of this Administration's disregard for and disrespect of science and scientists—and may signal the beginning of a disturbing trend to reduce the role of science in protecting the quality of the air we breathe.

Political interference in the PM case is clear. Incorporating changes and opinions from the White House Office of Management and Budget while ignoring the advice of its own scientists and the independent scientists on its Clean Air Scientific Advisory Committee (CASAC), the EPA set a National Ambient Air Quality Standard (NAAQS) for particulate matter that is *not* based on the best available science and does *not* protect the public's health with an adequate margin of safety, as required by the Clean Air Act. This is the first time in the CASAC's almost 30-year history that an EPA Administrator has disregarded its advice.

As the EPA moves on to consider acceptable levels of ozone pollution, there is concern that we may be in for a repeat performance. The CASAC recently and unanimously recommended a reduction in the NAAQS for ozone. A proposed standard is expected from the EPA as early as March 2007. Will EPA Administrator Stephen Johnson make it 0 for 2?

And more trouble may be in the wings. An EPA work group is reviewing the overall process for setting the NAAQS and has made some recommendations that should signal concern. Specifically, the work group report recommends that the EPA "Staff Paper"—a document prepared by agency scientists that identifies the most policy-relevant science and presents results from quantitative risk assessments—be replaced with a "narrowly focused policy assessment document" that should "reflect the agency's views."

The CASAC provided initial comments on the agency's recommendations and made some of its own. They include convening a science workshop as the first step in the five-year review of each NAAQS and the development of a Science Assessment document that would essentially include the Staff Paper and "would represent Agency staff's recommendation on the quality of current science relative to the question of adverse health and environmental effects at the existing air quality standard . . . [and] then defend that staff interpretation of the scientific literature through its summary of studies that directly address the critical, policy-relevant questions." This issue merits careful watching as it unfolds. [CASAC comments on the proposal can be viewed at: www.ucsusa.org/scientific_integrity/interference/naaqs-epa-staff-paper.html].

Unfortunately, the misuse of science extends far beyond air pollution. For several years, the Union of Concerned Scientists has been actively monitoring and documenting the misuse of science in public policy-making—from the stacking and use of political litmus tests on federal advisory committees to the blatant censorship and manipulation of agency science to efforts to muzzle federal scientists. The level and breadth of political interference with the integrity of science in our federal agencies is truly remarkable. The Administration has exerted political influence on a host of scientific issues, including global warming, reproductive health, endangered species, childhood lead poisoning, mercury emissions from power plants, condoms, and more. In the context of interfering with agency science, this Administration is truly an equal opportunity employer. [For more information, visit http://www.ucsusa.org/ scientific_integrity/interference/]

Red pens, silencers, and heavy hands are not the only instruments being used in this effort to put a lid on science that fails to support or comport with predetermined policies and ideological preferences. Although reported in an editorial in *Nature* (443, 121 (14 September 2006), one method that has flown surprisingly quietly under the radar screen is the repeal of whistleblower protection for federal workers under the Clean Water Act (CWA). Unlike the government-wide whistleblower protection law which offers scant protection and only in the reporting of a crime or gross mismanagement, the CWA was written to provide protection to public- and private-sector workers who report on failures of enforcement and/or clean-up—*and on the manipulation of science.* In 2005, the Department of Justice quietly removed that protection—previously available to some 170,000 members of the federal work force—by invoking the arcane legal doctrine of sovereign immunity. Adding insult to injury, the change was unpublished and unknown to affected federal workers for a year and only came to light when Public Employees for Environmental Responsibility (PEER) found a reference to the new policy in a footnote to a whistleblower complaint and obtained a copy of it under the Freedom of Information Act.

Political censorship, manipulation, and distortion of science and the muzzling of agency scientists hurt all of us. As with the case of the PM standard, the impacts on public health and the environment can be direct, acute, and long-term; other impacts are more subtle and insidious.

Politicized science erodes public trust in government and respect for science, tarnishing even honest, independent science and scientists and diminishing faith in agencies traditionally held in high esteem. This could make it difficult to attract and retain top notch scientists in our government agencies and—after decades of investment in these public resources—seriously erode our nation's capacity to address public health, safety, and environmental problems and to retain its scientific leadership. UCS surveys of scientists at the FDA, NOAA, and USFWS have documented the demoralizing impact of political interference on agency scientists.

Equally worrisome is the potential reluctance of independent scientists and professionals to serve on federal advisory committees and peer review panels. This has been a critical component of our democratic policy-making process. We need objective, transparent, and independent science to inform public policy.

While science and scientists aren't perfect, science is a cornerstone of public health, environmental sustainability, and economic growth. Although science will always be pressured by politics on some level, we need to set high federal standards that respect the scientific process and create reforms that prevent political leaders from misusing science. It won't happen on its own, and the problem won't go away with the advent of a new Administration. Although this Administration's pattern of interference has been remarkable, it's not the first nor will it be the last to politicize science.

So consider this a call to arms. Readers of this journal understand and care about the integrity and the outcomes of our public policy process. You work in the public interest. You or your colleagues participate in government-funded research or serve on federal advisory committees. You have credibility and influence: people will listen when you speak out about the importance of independent science to public health and well-being.

If you haven't done so already, now is the time to engage your elected officials on this issue. Demand they actively push legislation that makes it illegal to suppress or tamper with scientific information or release false documents; that provide meaningful whistleblower protection for scientists; and that provide a source of independent scientific advice for Congress and the President. Raise awareness among your friends, family, and colleagues about the real-world importance of this wonky-sounding issue. Engage your professional societies. Write op-eds. Be a source of support for your colleagues in federal agencies. Join us in keeping the issue visible and help us develop the solutions and reforms we can begin to implement in 2008.

The Union of Concerned Scientists has hit the ground running in a sustained initiative to restore scientific integrity to federal policy-making. We will give you the advice, information, and support necessary to make a meaningful contribution to this important effort. To learn more, and to show your support of independent science by joining 11,000 of your colleagues on a scientist statement on scientific integrity, visit www.ucsusa.org/rsi_statement.

Note: Originally published in *New Solutions, 17*(1-2), pp. 13-16, 2007.

Section II

PRECAUTIONARY SCIENCE

CHAPTER 6

PCBs in School—Persistent Chemicals, Persistent Problems

Robert F. Herrick

The issue of polychlorinated biphenyls (PCB) exposures resulting from occupancy of PCB-contaminated buildings is not new, but the contribution of building materials to that contamination is largely unrecognized. A rapidly emerging base of evidence shows that PCBs can be widely found in caulking and paint in masonry buildings constructed or renovated from about 1950 to the late 1970s. These materials can cause extensive PCB contamination of the building interiors and surrounding soil, and people who teach, live, or attend school in these buildings can have elevated serum PCB levels. The potential risk associated with this source of PCB exposure is not known; however, it is worth noting that the specific PCB congeners found at high levels in the building environments, and in biological samples from the occupants, include some that are suspected of being potent neurotoxins. The U.S. Environmental Protection Agency (EPA) is moving to address this issue in schools; however, the costs of remediating contaminated buildings will pose a formidable obstacle to most school districts.

The story of PCB contamination in schools is unfinished. We know that PCBs got into schools in building materials such as window caulking and paints. Given the time that has passed since these buildings were constructed, and the number of buildings that have already been renovated and had the PCB materials removed, the precise significance of the risk that PCBs from these sources pose to children, teachers, and other occupants of the schools may never be known. Although the presence of PCB caulk and paint in these buildings causes PCB contamination of the building interiors and the soil around the buildings, and is a clear violation of the federal "Toxic Substances Control Act" (40 CFR Part 761), the great majority of people who work, teach, or go to school in these buildings have no idea that they are being exposed to PCBs.

What is known is that PCBs can be present in buildings constructed or renovated from about the mid-1950s to the late 1970s, coinciding with a period of extensive school construction. Paints, the caulking material around window and door frames and in expansion joints and other building materials can contain very substantial amounts of PCB, sometimes in levels of hundreds of thousands of parts per million (ppm). For reference, the "Toxic Substances Control Act" (TSCA) prohibits material with PCB levels exceeding 50 ppm (40 CFR Part 761), and the U.S. Environmental Protection Agency considers this to be an "unauthorized use" of PCBs. It is also known that much of this material is at or near the end of its service life and is being replaced during building renovation projects, such as window replacements. In most cases, no precautions are taken to prevent the release of PCB-containing dust into the school environment while the old materials are scraped and ground out. This has resulted in extensive PCB contamination of building interiors and surrounding grounds. In other cases, the PCBs are being released from deteriorating caulk that is still in place but has weathered to the extent that the polymer is breaking down.

BACKGROUND

Indoor contamination with PCB is not new. It is very well documented that building environments can be contaminated as a result of transformer leaks, fires, and explosions. The EPA and states including California specifically addressed this source of PCB contamination, including some fact sheets targeted to schools [1, 2].

In the United States, the possibility of indoor PCB contamination from building materials such as caulk and paint went largely unrecognized, despite the attention paid to PCB releases from transformers and other sources. Starting in the early 1990s, investigators in Germany, Sweden, and Finland reported the presence of PCB in building sealants (caulking) and demonstrated relationships between these PCBs and elevated PCB levels in indoor air and settled dust [3]. Other investigators confirmed these findings and found extensive contamination of soil around the foundations of buildings containing these materials [4-8]. The EPA stated in 1996 that "very little is known about contaminated caulking and grout, their specific applications and dissemination in the marketplace. Samples of caulking that have been contaminated with PCBs have been found in a setting previously used as a school at a maximum concentration of 310,000 ppm PCBs" [9]. This reference is to a U.S. Army Corps of Engineers report, "Final Site Investigation Report for Campbell, Lyle, Stone and Otis Memorial Schools, Bourne, MA" [10]. In covering this story, the *Boston Globe* (March 21, 1996) reported that the Corps of Engineers sampled caulking, ceiling tiles, and paint in an elementary school on the Massachusetts Military Reservation and found PCB levels up to 2,000 times greater than allowed under state rules [11].

In the late 1990s, investigations in Sweden and Germany found that teachers and students in schools containing PCB caulking were exposed to elevated levels of PCB in the indoor air. These exposures caused significantly elevated serum levels of specific PCB congeners that were released from building materials [9, 12-14]. During a 2000 investigation conducted to address occupants' concerns about cancer and possible pesticide contamination in a university administration building [15, 16], PCB was detected in dust samples. The university closed the building, and subsequent testing revealed that gasket, foam board insulation, and caulking material contained PCB in quantities up to 33,000 mg/kg (ppm). In the case of this university building, EPA (Region 1, Boston), mandated a clean-up program to remove the identified PCB-containing materials and clean the building to clearance criteria of 1 microgram per cubic meter ($\mu g/m^3$) in air, 10 micrograms/100 square centimeters in surface wipes, and 1 ppm in porous building components (brick and concrete blocks). This abatement was mandated under the EPA finding that the PCB concentrations in caulk and other building materials exceeded 50 ppm, which constituted "unauthorized use" of non-liquid PCB products. The total cost of the abatement of the building exceeded $2 million.

THE CURRENT SITUATION

In 2003, E. Priha (Finnish Institute of Occupational Health) made a presentation at the American Industrial Hygiene Conference and Exposition describing a national project to remove PCB caulk from buildings in Finland [17]. One finding of that national initiative was the estimate by Finnish authorities that 220 tons of PCB had been used in over 10,000 buildings. As of 2003, approximately 85% of that material had been replaced with non-PCB containing caulking material.

Was the situation in the United States similar to that reported in Finland? To seek an answer, researchers at the Harvard School of Public Health called upon the support of the International Union of Bricklayers and Allied Craft Workers and one of their retired members. We had collaborated with this retiree in The Center for Construction Research and Training-supported research on dust controls. He had specialized in pointing, cleaning and caulking (PCC) work, and he remembered installing PCB-containing caulk in buildings throughout New England. He collected small (~ 10 g) samples of caulking from the exterior of 24 buildings. At the time of the sampling (2003), some buildings had already undergone renovation and caulking appeared to have been replaced. In others, reconstruction projects were underway, and, in some others, the original caulking was still in place but was obviously deteriorated and peeling away from the building joints. Of the 24 buildings sampled, 13 contained caulking material in which detectable levels of PCBs were measured, and eight of these 13 contained caulking that exceeded the 50 ppm EPA removal criteria by a factor of nearly 1,000 (range, 70.5–36,200 ppm; mean, 15,645 ppm) [18]. The elevated PCB levels in caulking were found in schools, university buildings, and other public buildings (Table 1).

This finding might have remained a scientific curiosity were it not for a concerned parent in Yorktown Heights, New York. Dr. Daniel Lefkowitz read the results of the sampling in Boston and wondered about his son's school which was built in 1969. Its windows had been replaced in 2003. He was curious to find out whether the caulking that had been removed in the window replacement contained PCBs, so he called the EPA Region 2. They advised him to look for a piece of caulking on the ground and send it to a lab for testing. He found a piece, had it tested (at his own expense), and got the result that the caulking contained 38,000 ppm PCB, 760 times the EPA's allowable limit.

Dr. Lefkowitz notified the Westchester County Department of Health of his finding, and they ordered the school to remove any remaining pieces of caulk on the ground. They refused Lefkowitz's request that the soil around the school be tested, so again he collected a sample and paid for the analysis himself. The soil sample came back with elevated levels of PCB above the New York State 1 ppm clean-up requirement. The Westchester Department of Health did its own soil testing and confirmed Lefkowitz's result of soil contamination, finding levels as high as 40 ppm (see Figure 1).

They ordered the school to fence off all of the contaminated areas and removed nearly 600 cubic yards of contaminated soil. The total cost of decontaminating the school was approximately $300,000, although the PCB caulking remaining in building expansion joints and around doors, etc was encapsulated in paint, and not removed until 2007.

Lefkowitz's findings and concerns were met with some skepticism, but he persisted, and created a website www.pcbinschools.org to disseminate information on PCB contamination in schools. The Westchester County government sought the advice of the EPA, and in response to a letter from Michael Kaplowitz, Chairman of Budget and Appropriations Committee of the Westchester County Board of Legislators, Kenneth S. Stoller, the Chief of Pesticides and Toxic Substances Branch, EPA Region 2 stated:

> The use of PCBs in caulking and sealant materials has never been authorized by the Environmental Protection Agency. In general, the placement of such materials pre-dates the enactment of the Toxic Substances Control Act and its use today is not authorized. Therefore, the prospect of authorizing the continued use of this material in residential settings, or where children could be exposed, is extremely unlikely. Because it is illegal, and the potential for exposure may be significant, PCB-containing caulk must be removed upon discovery [19].

Table 1. Sampling Results from 24 Buildings in Greater Boston

Building type	PCB Content (ppm by mass)[a]
Government office, mixed use	35,600
Government office	25.2
Office building	Not detected[b]
Subsidized housing	Not detected
Subsidized housing	Not detected
Elderly housing	Not detected
Elderly housing	Not detected
University student housing	36,200
University dormitory	70.5
University dormitory	1.68
University classrooms and offices	26,400
Elementary school	7,740
Middle school	5,010
High school	5,970
High school	Not detected
High school	Not detected
Community college	19.3
Church offices	2.14
Synagogue	8,240
Hospital	Not detected
Hospital	Not detected
Museum	0.56
Hotel	Not detected
Police station	Not detected

[a]All samples, except for the sample from the governmental office/mixed use building, were identified as Aroclor 1254. This sample was identified as Aroclor 1260.

[b]Samples reported as "Not detected" contained PCB at less than the reporting limit, which ranged from 114-455 µg/kg (0.114-0.455 ppm).

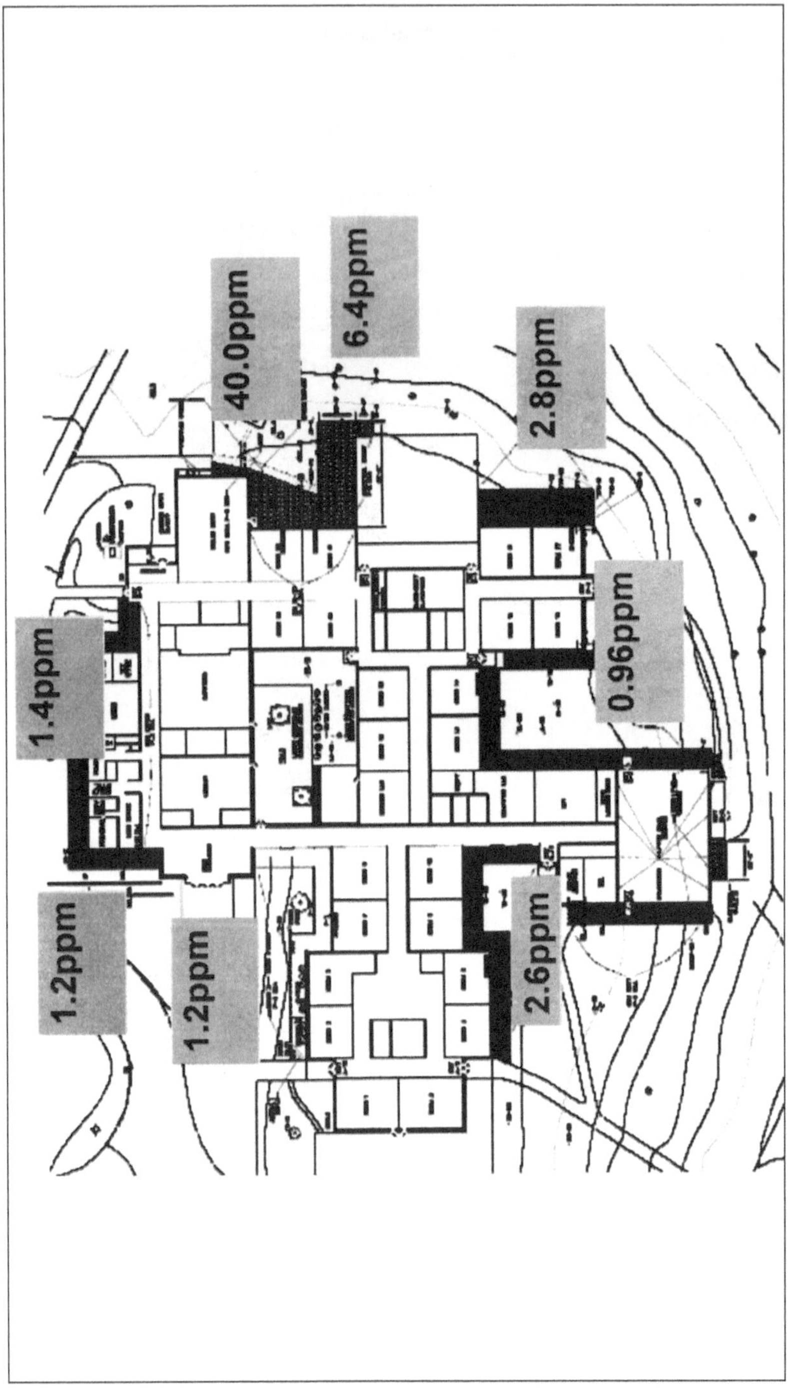

Figure 1. PCB soil contamination, French Hill School, Yorktown Heights, New York.

THE CONUNDRUM

The EPA's interpretation of TSCA is that if PCB in caulking (or presumably other building materials) is found at levels exceeding 50 ppm, it must be removed. This is clear from the 2005 letter to Kaplowitz and the EPA Fact Sheet entitled, "Current Best Practices for PCBs in Caulk (updated June 2009)" [20], which states that "caulk containing PCBs at levels > 50 ppm is not authorized for use under the PCB regulations and must be removed." In the case of the university administration building described earlier, the cost of abatement of a 10-story building was approximately $2 million. And while the regulations are clear about the need for removal of PCB-contaminated material upon discovery, there is no requirement that testing be done to find such materials. So what would motivate a building owner to test for PCB contamination in the first place?

So upon this conundrum, things sat for months. On the scientific side, research continued, with another study that showed that PCB soil contamination was possible even when caulking material was intact and undisturbed, and that PCB could be readily mobilized from old caulk by natural weathering [21]. Dr. Lefkowitz continued to push for action to remove the PCB caulk from his son's school and to broaden the investigation to similar buildings.

Dr. Lefkowitz also reached out to a reporter at the *New York Daily News*, Bill Egbert, following which the paper decided to sample caulk at a limited number of schools in New York City. The *Daily News* obtained samples from nine schools and sent them to a state-certified laboratory. Six of the nine samples contained more than the EPA's threshold proportion of 50 ppm PCBs—indeed, at two schools, the levels were literally thousands of times higher: 225,000 at PS 199 in Manhattan and 111,000 ppm at PS 178 in the Bronx. The paper ran a series of stories discussing the discovery, the community response, and a subsequent City Council hearing devoted to the issue. The New York City Department of Education responded to the concern expressed by parents and political leaders by conducting a series of widely criticized tests in the schools in the *Daily News* study and by minimizing the risk of exposure [22].

The *Daily News* series prompted two legal developments. The first one was legislative: New York State Assemblywoman Linda Rosenthal, whose district includes PS 199, the school where the caulk had tested highest for PCBs, sponsored a bill that would mandate testing of the caulk at all New York City schools built during the period when PCBs were in use as plasticizers. The bill garnered the support of a number of co-sponsors but was stalled during the chaos that characterized the New York State Legislature in summer 2009. It will be reintroduced during the next session.

In addition, New York Lawyers for the Public Interest (NYLPI), a civil rights non-profit with an environmental justice section, started a campaign around the issue with two components, community organizing and a litigation. NYLPI filed a formal notice of intent to sue New York City on behalf of Bronx mother Naomi

Gonzalez in March 2009; after the notice was filed, the City started meeting with the EPA to discuss the problem of PCB-contaminated caulk. In January 2010, several months after the lawsuit itself was filed in federal court, the City reached agreements with both the EPA and the plaintiff. It will conduct a pilot study to assess the problem's scope and potential remedies, ultimately generating a citywide plan. On a parallel track, in September 2009 EPA headquarters announced new national guidances on PCBs in schools and two major research initiatives.

LABOR'S ROLE

The investigation in the United States would never have progressed without the support of labor organizations. The retired bricklayer who identified potentially contaminated buildings in Boston was trained in industrial hygiene principles as part of a program at the CPWR (formerly the Center to Protect Workers' Rights, now the Center for Construction Research and Training). The initial survey that discovered elevated PCB caulking levels in eight of 24 buildings was supported by the International Union of Bricklayers and Allied Craft Workers, Local 3, Eastern Massachusetts. Construction trade workers participated in the first study done in the United States that demonstrated substantial differences in PCB serum levels between construction workers and general population samples. These findings strongly suggest that occupational contact with caulking material can be a major source of PCB exposure for construction workers [23].

More recently, the Massachusetts Teachers Association (MTA) has expressed interest in evaluating the extent to which schools may contain PCB building materials and the possible effect these PCBs may have on their members. The MTA and investigators from Harvard School of Public Health are addressing these questions.

THE WAY FORWARD

Until it posted a Web site on PCB in building caulk in 2008 [24], the U.S. EPA appeared to be in a state of vincible ignorance on the subject. Although PCBs are the only chemicals specifically mentioned in the "Toxic Substances Control Act," and in 1996 the EPA clearly was aware that building caulk could contaminate the interior of buildings [10], the agency seemed willfully ignorant of the problem. In fact, a telephone survey of the 10 EPA regions conducted in 2007 showed a great deal of variability between regions. Regions 1 (Boston), 2 (New York), and 8 (Denver) reported activity on PCB in buildings and other structures, while some others seemed to have little awareness that PCB in building materials was an issue [25]. With the arrival of new leaders in 2009, the EPA has acted to develop guidelines on caulking as a source of PCBs in schools. This guidance represents a promising start, while the effect it will have in practice is yet to be seen. In its September 25, 2009 press release, the EPA stated that ". . . unfortunately high levels of PCBs are present in many buildings and facilities constructed prior to the PCB ban, including most recently some schools. We're concerned about the potential risks associated with exposure to these PCBs . . ." [26]; however, the underlying conundrum in the TSCA regulation remains: There is no requirement that a building owner test for PCB in caulking. But if the owner or another concerned party such as a labor organization or a parent tests for PCB and finds it, TSCA requires that it be removed. In the case of Berkshire (Massachusetts) Community College that did the right thing and tested for PCB in caulk before starting a renovation project, the cost of the project more than doubled from $2.2M to an estimated $5M [27].

Can cash-strapped school districts reasonably be expected to take on the economic burden of addressing PCBs in their buildings? In some cases, renovations that are already planned could easily double in cost, but, in many school districts, needed renovations such as window replacements have been deferred for years. Our data suggests that if testing is done, approximately one-third of the schools constructed from about 1950 to the late 1970s will be found to contain PCBs. To comply with TSCA, these PCBs must be removed or the schools could risk severe penalties, should the EPA pursue enforcement. The only reasonable course to protect teachers, staff, and students from what the EPA acknowledges is a source of concern is to provide federal financial support for PCB testing and removal. House Bill 2187, the "21st Century Green High-Performing Public School Facilities Act" providing such support, was passed on May 18, 2009 by the U.S. House of Representatives and sent to the U.S. Senate Committee on Health, Education, Labor and Pensions. This legislation provides "grants . . . for the purpose of modernizing, renovating, or repairing public school facilities, based on their need for such improvements, to be safe, healthy, high-performing, and up-to-date technologically" [28]. Under allowable uses of funds, this legislation specifies that the grants shall be used for modernization, renovation, or repair of public school facilities, including, where applicable, early learning facilities, including "abatement, removal, or interim controls of asbestos, polychlorinated biphenyls, mold, mildew, or lead based hazards, including lead-based paint hazards."

CONCLUSIONS

This story has no conclusion; however, there are several undisputable facts:

- Masonry buildings constructed or renovated between about 1950 and the late 1970s may contain PCBs in materials such as caulking and paint, sometimes in very high concentrations.
- PCBs are being released from these materials, contaminating the building interiors and the soil around these buildings.

- Occupants of these buildings may have significantly elevated serum PCB levels, particularly for the more volatile PCB congeners.
- The potential risk from this source of PCB exposure has not been determined; however, there is growing evidence that some of the more volatile PCB congeners released from caulk can be potent neurotoxins [29].
- The EPA has signaled that they intend to strengthen chemical management laws by making changes in TSCA. Specifically, they have identified initiating rulemaking under section 6 of TSCA to re-evaluate the TSCA PCB use and distribution in commerce regulations as a topic for action [30]. This offers the hope that they will develop a rational public policy to minimize PCB exposure from what appears to be a very significant source.

ACKNOWLEDGMENTS

The contributions of Miranda K. S. Massie, Senior Staff Attorney, New York Lawyers for the Public Interest, are gratefully acknowledged.

REFERENCES

1. U.S. Environmental Protection Agency, *PCBs in Fluorescent Light Fixtures,* http://yosemite.epa.gov/R10/EXTAFF.NSF/84c2eal918cf8b90882566e200738f51/8078cb9ed319d3ed882568870002645e/$FILE/pcbfacts.pdf accessed September 30, 2009.
2. California Department of Toxic Substances Control, *Fact Sheet—PCBs in Schools,* http://www.dtsc.ca.gov/Schools/upload/SM_FS_PCB_Schools.pdf accessed September 30, 2009.
3. C. Benthe, B. Heinzow, H. Jessen, and W. Rotard, Polychlorinated Biphenyls. Indoor Air Contamination Due to Thiokol-Rubber Sealants in an Office Building, *Chemosphere,* pp. 1481-1486, 1992.
4. E. Balfanz, J. Fuchs, and H. Kieper, Sampling and Analysis of Polychlorinated Biphenyls (PCB) in Indoor Air due to Permanently Elastic Sealants, *Chemosphere, 26,* pp. 871-880, 1993.
5. U. Burkhardt, et al., Indoor Air Pollution by Polychlorinated Biphenyl Compounds in Permanently Elastic Sealants, *Offentl Gesundheitweis, 52,* pp. 567-574, 1990.
6. R. Corner, et al., PCB in Indoor Air and Dust in Buildings in Stockholm, in *Indoor Air 2002*, proceedings from the 9th International Conference on Indoor Air Quality and Climate, Monterey, California, *4,* pp. 141-146, June 30–July 5, 2002.
7. H. Fromme, et al., Polychlorinated Biphenyls (PCB) in Caulking Compounds of Buildings—Assessment of Current Status in Berlin and New Indoor Air Sources, *Gesundheitswesen, 58,* pp. 666-672, 1996.
8. V. Pyy, and O. Lyly, *PCB in Jointing Materials in Prefabricated Houses and in the Soil of Yards* (Helsinki, Finland: City of Helsinki Environment Centre, 1998).
9. T. Gabrio, et al., PCB-Blood Levels in Teachers, Working in PCB-Contaminated Schools, *Chemosphere, 40:*9-11, pp. 1055-1062, 2000.

10. Use Authorization for, and Distribution in Commerce of, Non-liquid Polychlorinated Biphenyls; Notice of Availability; Partial Reopening of Comment Period, *Federal Register, 64,* 237 (December 10, 1999.
11. S. Leung, Source of Toxin Revealed at Bourne School, *Boston Globe*, March 21, 1996.
12. F. Neisel, et al., Human Biomonitoring of Polychlorinated Biphenyls in 130 Exposed Elementary School Children, *Gesundheitswesen, 61,* pp. 137-149, 1999.
13. M. Schwenk, et al., Human Biomonitoring of Polychlorinated Biphenyls and Polychlorinated Dibenzodioxins and Dibenzofuranes in Teachers Working in a PCB Contaminated School, *Chemosphere, 47,* pp. 229-233, 2002.
14. B. Liebel, et al., Evidence for Increased Internal Exposure to Lower Chlorinated Polychlorinated Biphenyls (PCB) in Pupils Attending a Contaminated School, *International Journal of Hygiene and Environmental Health, 237,* pp. 315-324, 2004.
15. K. Coghlan, et al., Characterization of Polychlorinated Biphenyls in Building Materials and Exposures in the Indoor Environment, in *Indoor Air 2002*, proceedings from the 9th International Conference on Indoor Air Quality and Climate, Monterey, California, June 30–July 5 (2002) 4, 147-152.
16. M. Chang, K. Coghlan, and J. McCarthy, Remediating PCB-Containing Building Products: Strategies and Regulatory Considerations, in *Indoor Air 2002*, proceedings from the 9th International Conference on Indoor Air Quality and Climate, Monterey, California, *4,* pp. 171-176, June 30–July 5, 2002.
17. E. Priha, Health and Environmental Aspects of PCB Contamination Due to Old Polysulfide Sealants, from podium session 104, presentation no. 25, American Industrial Hygiene Conference and Exposition, Dallas, Texas (May 2003).
18. R. F. Herrick, M. D. McClean, J. D. Meeker, L. K. Baxter, and G. A. Weymouth, Commentary: An Unrecognized Source of PCB Contamination in Schools and Other Buildings, *Environmental Health Perspective, 112,* pp. 1051-1053, 2004. PMC1247375.
19. U.S. Environmental Protection Agency, Letter to Michael Kaplowitz, November 3, 2005, http://www.pcbinschools.org/EPA%20LETTER.pdf accessed September 30, 2009.
20. U.S. Environmental Protection Agency, Current Best Practices for PCBs in Caulk Fact Sheet: Removal and Clean-Up of PCBs in Caulk and PCB-Contaminated Soil and Building Material, http://www.epa.gov/pcbsincaulk/caulkremoval.htm accessed September 30, 2009.
21. R. F. Herrick, D. J. Lefkowitz, and G. A. Weymouth, Commentary: Soil Contamination from PCB-Containing Buildings, *Environmental Health Perspective, 115,* pp. 173-175, 2007.
22. Testimony of Kathleen Grimm, Deputy Chancellor, Finance and Administration, New York City Department of Education, New York City Council Committees on Education, Environmental Protection, and Oversight and Investigations Special Hearing on PCBs in Schools, April 29, 2008.
23. R. F. Herrick, J. D. Meeker, R. Hauser, L. Altshul, and G. A. Weymouth, Serum PCB Levels and Congener Profiles among US Construction Workers, *Environmental Health, 6,* p. 25, 2007. PCM2034547.
24. U.S. Environmental Protection Agency, PCBs in Caulk in Older Buildings, http://www.epa.gov/waste/hazard/tsd/pcbs/pubs/caulk/index.htm accessed September 30, 2009.

25. M. Parks, research report submitted in partial fulfillment of M.S. degree requirements, Harvard School of Public Health, May, 2007.
26. U.S. Environmental Protection Agency, *EPA Announces Guidance to Communities on PCBs in Caulk of Buildings Constructed or Renovated Between 1950 and 1978: EPA to Gather Latest Science on PCBs in Caulk,* http://yosemite.epa.gov/opa/admpress.nsf/d0cf6618525a9efb85257359003fb69d/28c8384eea0e67ed8525763c0059342f!OpenDocument accessed September 30, 2009.
27. J. Smith, Berkshire Community College Cleanup to Cost $5M, *Berkshire Eagle*, September 5, 2009.
28. *21st Century Green High-Performing Public School Facilities Act*, HR 2187, http://thomas.loc.gov/cgi-bin/bdquery/z?d111:HR02187 accessed September 30, 2009.
29. T. Simon, J. Britt, and R. James, Development of a Neurotoxic Equivalence Scheme of Relative Potency for Assessing the Risk of PCB Mixtures, *Regulatory Toxicology and Pharmacology, 48,* pp. 148-170, 2007.
30. U.S. Environmental Protection Agency, *Enhancing EPA's Chemical Management Program,* http://www.epa.gov/oppt/existingchemicals/pubs/enhanchems.html accessed September 30, 2009.

Note: Originally published in *New Solutions, 20*(1), pp. 115-126, 2010.

CHAPTER 7

Chrysotile Asbestos Exposure: Cancer and Lung Disease Risks

John M. Dement

Chrysotile is the predominant type of asbestos used in the United States and most other countries historically, and thus represents the most important source of exposure to asbestos already in place. Chrysotile use in developing countries is increasing. While all types of asbestos are regulated equally in the United States [1, 2], some continue to maintain that chrysotile is less carcinogenic than other asbestos types [3]. Supporting the strict regulation of chrysotile are studies of textile workers exposed to chrysotile or predominantly chrysotile which have consistently shown excess lung cancer and pneumoconiosis. These cohorts also have demonstrated steep increases in the risk of lung cancer and pneumoconiosis with cumulative asbestos exposure [4-8].

In previous publications, exposure and mortality of workers in a South Carolina textile plant processing chrysotile asbestos were reported [4, 5, 9]. In order to further refine the risk estimates associated with chrysotile exposure, the South Carolina cohort has been updated and reanalyzed by adding 15 years of observation [10].

STUDY POPULATION AND METHODS

Detailed descriptions of the plant studied as well as methods used for the update of this cohort have been published elsewhere [4, 5, 9, 10]. Briefly, the plant was located in South Carolina and began production of asbestos products in 1896. Asbestos textiles were first produced in 1909. Chrysotile, received from Quebec, British Columbia, and Rhodesia, was the only type of asbestos processed as a raw fiber. This plant in unique in that most major engineering controls to reduce asbestos exposures were in place by 1940, and the plant processes remained fairly constant during the period from 1940 until production of asbestos textile closed in the late 1970s. Chrysotile exposure levels (expressed as fibers longer than 5 micrometer per cubic centimeter of air) by areas of the plant (department and operations), specific textile jobs, and calendar years were estimated based on historic data and were used in the previously published analyses [5].

In the original analysis, the study cohort was defined as all white male workers employed in the textile production operations for at least one month between January 1, 1940 and December 31, 1965 with cohort follow-up through December 31, 1975. This presentation is concerned with white males who met the same employment requirements with follow-up of vital status extended from 1975 to December 31, 1990. All of the basic data coded for the original study, including the work histories at the plant and follow-up information through 1975, were used in the update. Traditional sources were used for determining vital status of workers and obtaining death certificates. The life table analysis system (LTAS) of the National Institute for Occupational Safety and Health [11] was used to calculate the standardized mortality ratios (SMRs) for specific causes of death. Age, race, sex, and calendar time specific death rates for the U.S. population were used to calculate expected deaths and SMRs. Full detail of study methods has been published elsewhere [10].

RESULTS AND DISCUSSION

Results of vital status update are given in Table 1. Among white males, the number of deaths almost doubled from 308 to 607 and an additional 11,000 Person Years at Risk (PYAR, from 33,141 to 44,131) were included in the update. Table 2

Table 1. Vital Status of Asbestos Textile Cohort

Vital status	White males
Alive	621 (49.8%)
Dead	607 (48.7%)
Unknown	19 (1.5%)
Total	1247*
PYAR	44,131
Avg. years of observation	35

Notes: Death certificates were missing for 17 members of the cohort; PYAR = Person Years At Risk.

Table 2. Mortality by Major Cause of Death Categories for the Asbestos Textile Cohort Based on U.S. Death Rates

Cause of death category (9th Revision ICD codes)	White males	
	Observed	SMR A
Tuberculosis (010–018)	6	1.63 (0.71–3.22)
All cancer (140–208)	137	1.46 (1.26–1.68)
Diabetes mellitus (250)	12	2.02 (1.17–3.28)
Heart disease (390–398, 402, 404, 420–429)	226	1.37 (1.23–1.53)
Cerebrovascular disease (430–438)	31	1.46 (1.05–1.96)
Respiratory disease (460–466, 470–478, 480–487, 490–519)	58	2.24 (1.78–2.79)
Pneumoconiosis and other respiratory (470–478, 494–519)	41	3.97 (3.01–5.13)
Digestive disease (520–577)	23	1.07 (0.73–1.52)
Accidents (E800–E949)	47	1.47 (1.13–1.87)
All other and unknown	67	1.27
All causes	607	1.44 (1.35–1.54)

Notes: ICD = International Classification of Diseases; A = Standardized mortality ratio (SMR); numbers in parentheses are 90% confidence intervals. Lost-to-follow-up considered alive in these analyses.

gives the results of the mortality analysis for major causes of death based on the use of U.S. mortality rates to generate the expected deaths. The mortality for all causes combined (SMR = 1.44; CI = 1.35–1.54) and for several of the major causes of death are greater than expected. The excesses observed for cancer (SMR = 1.46; CI = 1.26–1.68), diabetes mellitus (SMR = 2.02; CI = 1.17–3.28), heart disease (SMR = 1.37; CI = 1.23–1.53), cerebrovascular disease (SMR = 1.46; CI = 1.05–1.96, non-malignant respiratory disease (SMR = 2.24; CI = 1.78–2.79), and accidents (SMR = 1.47; CI = 1.13–1.87) are all statistically significant. The mortality in the category non-malignant respiratory diseases is primarily due to pneumoconiosis (asbestosis) and "other respiratory disease" (primarily chronic obstructive pulmonary disease). Pneumoconiosis is the underlying cause of death for approximately 50 percent (21/41) of the deaths in this category.

Table 3 gives the mortality results for malignant neoplasms of specific sites. A statistically significant increase in all respiratory tract cancer (SMR = 2.25; CI = 1.85–2.71), primarily due to lung cancer, and an unexpected statistically significant increase in liver/gall bladder/biliary tract cancer (SMR = 4.04; CI = 1.76–7.97) was observed. Cancers of the pancreas, gastrointestinal tract, kidney, and larynx were observed in excess, although these excesses were not statistically significant. Mesothelioma (one pleural and one peritoneal) was recorded on the death certificates of two white males who were employed in the plant for 25 and 32 years and who had latency periods of 37 and 34 years, respectively. One death due to "cancer of the pleura" was observed without

Table 3. Mortality from Selected Malignant Neoplasms for the Asbestos Textile Cohort Based on U.S. Rates

Cause of death category (9th Revision ICD codes)	White males	
	Observed	SMR A
All digestive and peritoneum (150–159)	23	0.94 (0.63–1.34)
Stomach (151)	3	0.75 (0.21–1.95)
Liver, gall bladder, biliary (155.0, 155.1,156)	6	4.04 (1.76–7.97)
Pancreas (157)	7	1.46 (0.68–2.74)
All respiratory (160–165)	78	2.25 (1.85–2.71)
Larynx (161)	3	2.24 (0.61–5.79)
Trachea, Bronchus, Lung (162, 163.0)	74	2.24 (1.83–2.72)
Other respiratory (160, 163.1, 163.9)	1	2.58 (0.13–12.2)
Lymphatic and hematopoietic (200–208)	9	1.01 (0.53–1.77)
All other	27	1.04
All cancer (140–208)	137	1.46 (1.26–1.68)

Notes: ICD = International Classification of Diseases; A = Standardized mortality ratio (SMR); numbers in parentheses are 90% confidence intervals. Lost-to-follow-up considered alive in these analyses.

Table 4. Lung Cancer Mortality by Cumulative Exposure among the Asbestos Textile Cohort with at Least 15 Years Latency

Cumulative exposure (fibers/cc-days)	< 500	500 1,000	1,000 2,500	2,500 10,000	10,000 40,000	40,000 100,000	Total
Observed/ Expected	7/7.6	4/5.5	15/6.2	10/5.1	16/5.2	18/2.2	72/32.0
SMR	0.92	0.73	2.42*	1.96**	3.08*	10.00**	2.25*

*$p < 0.01$, **$p < 0.05$.

mention of mesothelioma on the death certificate. An additional mesothelioma occurring after 1990 has been confirmed in a white male employed in mostly non-textile operations at this plant. He was not included in the cohort.

Table 4 gives the lung cancer mortality by cumulative exposure as measured in fibers/cc-days. There is a strong and consistent increase in the risk of lung cancer with increasing cumulative exposure. A statistically significant increase in risk is first observed at a cumulative exposure of 1,000–2,500 fibers/cc-days and increases to an SMR of 10 among those with exposure in excess of 100,000 fibers/cc-days.

Results of the updated analyses of the South Carolina cohort confirm earlier observations of a steep lung cancer exposure-response relationship among chrysotile asbestos textile workers. Results obtained from the South Carolina cohort are consistent with those from other textile worker cohorts exposed to chrysotile or predominantly chrysotile [6-8].

Exposure to chrysotile will continue to be a major health concern for workers as well as the general public due to the millions of tons already in our factories, schools, and other buildings. Chrysotile use in developing countries is increasing. Serious campaigns have been launched by the chrysotile asbestos mining industry to minimize the hazards of chrysotile. Available epidemiological and animal bioassay data provide no scientific basis for differential public health policy with respect to chrysotile.

REFERENCES

1. Occupational Safety and Health Administration, Code of Federal Regulations, 29 CFR Parts 1910 and 1926, Occupational Exposure to Asbestos, Tremolite, Anthophyllite, and Actinolite; Final Rules, *Federal Register, 51,* pp. 22612-22790, 1986.
2. Environmental Protection Agency, *Airborne Asbestos Health Assessment Update,* Environmental Protection Agency, Washington, D.C., 1986.

3. B. T. Mossman, J. Bignon, M. Corn, A. Seaton, and J. B. L. Gee, Asbestos: Scientific Developments and Implications for Public Policy, *Science, 247,* pp. 294-301, 1990.
4. J. M. Dement, R. L. Harris, M. J. Symons, and C. M. Shy, Estimates of Dose-Response for Respiratory Cancer Among Chrysotile Asbestos Textile Workers, *Annals of Occupational Hygiene, 26,* pp. 869-887, 1981.
5. J. M. Dement, R. L. Harris, M. J. Symons, and C. M. Shy, Exposures and Mortality among Chrysotile Asbestos Workers. Part II: Mortality, *American Journal of Industrial Medicine, 4,* pp. 421-433, 1983.
6. A. D. McDonald, J. S. Fry, A. J. Woolley, and J. C. McDonald, Dust Exposure and mortality in an American Chrysotile Textile Plant, *British Journal of Industrial Medicine, 40,* pp. 361-367, 1983.
7. A. D. McDonald, J. S. Fry, A. J. Woolley, and J. C. McDonald, Dust Exposure and Mortality in an American Factory Using Chrysotile Amosite, and Crocidolite in Mainly Textile Manufacturing, *British Journal of Industrial Medicine, 39,* pp. 368-374, 1983.
8. J. Peto, R. Doll, C. Hermon, W. Binns, R. Clayton, and T. Goffe, Relationship of Mortality to Measures of Environmental Asbestos Pollution in an Asbestos Textile Factory, *Annals of Occupational Hygiene, 29,* pp. 305-355, 1985.
9. J. M. Dement, R. L. Harris, M. J. Symons, and C. M. Shy, Exposures and Mortality among Chrysotile Asbestos Workers. Part I: Exposure Estimates, *American Journal of Industrial Medicine, 4,* pp. 399-419, 1983.
10. J. M. Dement and D. P. Brown, Cohort Mortality and Case-Control Studies of White Male Asbestos Textile Workers, *Journal of Occupational Medicine and Toxicity, 2,* pp. 355-363, 1993.
11. K. Steenland, J. Beaumont, et al., New Developments in the Life Table Analysis System of the National Institute for Occupational Safety and Health, *Journal of Occupational Medicine, 32,* pp. 1091-1098, 1998.

Note: Originally published in *New Solutions* (Summer) pp. 5-8, 1994.

CHAPTER 8

Manganese in Gasoline: Are We Repeating History?

Gina M. Solomon, Annette M. Huddle,
Ellen K. Silbergeld, And Joseph Herman

In 1995, tetraethyl lead was finally phased out of gasoline in the United States. That same year, however, the Ethyl Corporation began selling another metal-based gasoline additive in this country. Ethyl is the same company which previously made and sold tetraethyl lead (TEL). The new additive, methylcyclopentadienyl manganese tricarbonyl (MMT) is used for the same purpose as TEL: it enhances octane and thereby acts as an anti-knock agent in gasoline. The Environmental Protection Agency (EPA) had refused for years to allow Ethyl to market MMT for addition to unleaded gasoline in the U.S. on the grounds that "there is a reasonable basis for concern about the effects on public health that could result (from manganese in gasoline)" [1]. The company challenged this EPA ruling in court and won. The court of appeals held that the EPA had exceeded its authority under the Clean Air Act (CAA), and directed the agency to grant the necessary license. Thus, the court ruling has removed any barrier to introduction of this additive into our fuel supply.

Manganese, unlike lead, is an essential trace element and is usually present in the human diet. In addition, predicted ambient levels of manganese are expected to remain fairly low despite inclusion of manganese in fuel. Nonetheless, major concerns remain. Inhalation is an abnormal route of intake for manganese and may be associated with increased risk of toxicity, particularly to the central nervous system and the lungs. Certain susceptible subpopulations, such as the young, the old, and the malnourished, may be at greatly increased risk of adverse effects from exposure to manganese in the environment. As is the case with lead, manganese may accumulate over time in the environment and in the body leading to potential overexposures. There are also significant occupational health concerns for workers in the mining, manufacturing, and blending stages. Although less concentrated, exposures in gas stations and garages may also lead to significant health sequelae. Finally, MMT may damage emission control systems in automobiles, thereby potentially leading to increased air pollution and increased repair costs to consumers. Given that there are viable alternatives to MMT use, it is unwise to proceed with the marketing and addition of MMT into the fuel supply of the United States.

HISTORY AND PRODUCTION

MMT was originally developed in the 1920s and was introduced in 1957 as an experimental anti-knock agent for gasoline. The developer, Ethyl Corporation, was founded to produce and market tetraethyl lead (TEL). MMT was originally designed to be added to leaded gasoline, but with the phase-out of TEL, MMT was reevaluated as an additive to unleaded fuel. In the period from 1977 to 1979, MMT was added to unleaded fuel on a temporary basis to increase fuel efficiency during the gasoline shortage. In 1978 Ethyl applied to the EPA for permission to market this chemical as a fuel additive in unleaded gasoline. The Clean Air Act requires that no new gasoline additives may be introduced without permission, in the form of a waiver, from the EPA; the waiver may only be granted if the additive does not interfere with the catalytic converter and thereby contribute to increased air pollution. Permission was denied at that time on the basis that MMT "had a statistically significant adverse HC [hydrocarbon! emissions effect" [1]. Ethyl then spent 10 years trying to show that MMT has no adverse effects on catalytic converters. EPA in 1994 acknowledged that MMT may not damage catalytic converters, but then refused to license the additive due to the potential health effects of manganese accumulation in the environment. It was this latter decision that was recently overturned by the courts.

Over the last 20 years, MMT has been produced solely by Ethyl Corporation at a plant in Orangeburg, South Carolina, and has been sold as a fuel additive. It has been added to virtually all unleaded gasoline in Canada since 1977. In addition, it has been sold as a fuel oil combustion improver for boilers, and for use in military aircraft as a fuel additive to decrease "telltale exhaust" [2]. This latter use, however, is decreasing due to the more efficient design of newer jet engine turbines. In 1983, Ethyl Corporation reported manufacturing approximately four million pounds of MMT annually, of which about 100,000 pounds were marketed as CI-2™ additive for boiler and turbine fuels, two million pounds were sold as an additive for leaded gasoline domestically (TEL Motor 33 Mix™), and the rest was exported as HiTEC 3000™ to be added to unleaded gasoline [3].

LEGAL REASONING

The CAA generally prohibits introduction of new fuel additives which are not "substantially similar" to existing fuels or fuel additives. Under section 211(f)(4) of the CAA, however, the EPA may grant a waiver allowing marketing of a new additive if the manufacturer can show that it "will not cause or contribute to a failure of any emission control device or system" [4], once on the market, section 211(c) applies. This allows the administrator of the EPA to "control or prohibit the manufacture or sale of any [currently used] fuel additive if she determines that 'any emission product of such . . . fuel additive causes, or contributes, to air pollution which may reasonably be expected to endanger the public health or welfare" [5]. Thus, the "public health or welfare" is explicitly considered only in action to remove already marketed fuel additives. In any action to remove a fuel additive from the market, the burden of proof falls on the EPA to show that the additive is unsafe. This is in marked contrast to section 211(f)(4) of the CAA where the burden of proof is on the manufacturer to show that its proposed additive is safe.

Ethyl Corporation applied for a waiver under section 211(f)(4) to allow marketing of MMT in unleaded fuel, first in 1978, and then again in 1981 and 1991. MMT was already in use at that time in leaded fuel and in jet fuel. EPA denied all three applications for a waiver on the grounds that Ethyl failed to demonstrate that MMT would not contribute to the failure of emission control systems. Finally, in 1994, after threat of litigation and provision of additional test data on the effects of MMT on automotive emissions, EPA determined that "Ethyl's product HiTEC 3000 in unleaded gasoline at the specified concentration will not cause or contribute to a failure to achieve compliance with vehicle emission standards" [5]. The EPA still refused the waiver, however, on the grounds that "there is a reasonable basis for concern about the effects on public health that could result if EPA were to approve use of MMT" [5].

Ethyl challenged the EPA decision in court and, in April of 1995, the D.C. Circuit Court of Appeals decided that the EPA may only consider emission effects and not public health effects in deciding whether or not to grant a waiver under section 211(f)(4) of the CAA. The court stated that "The Administrator plainly oversteps her authority when she considers other factors 'in the public interest' such as the public health implications of an additive in . . . waiver proceedings" [5]. In a further setback, the court decided that the EPA rules requiring pre-market testing of fuel additives (promulgated in May of 1994) did not apply to MMT, as the application for a waiver was filed prior to that date. The court therefore forced the EPA to grant the waiver and to make it retroactive to November 1993 [6]. Thus, the court ruled that EPA was not authorized to require that Ethyl conduct more extensive health testing of MMT prior to marketing. EPA is requiring further health testing of the fuel additive, but the product is already on the market.

BIOLOGICAL FATE

MMT is an organic manganese compound which decomposes rapidly in sunlight. It is extremely hazardous by all routes of exposure, and is well absorbed through the skin. MMT concentrates in the lungs, where it causes severe damage in experimental animals [7]. It is also a confirmed genotoxic agent [8]. When MMT is used in automotive fuel, breakdown products emitted from the tailpipe consist of a variety of inorganic manganese oxides, primarily Mn_3O_4 [9]. Essentially all manganese particulate emitted from the tailpipe is small enough to be respirable [9].

In the adult, much of an oral manganese dose does not cross the intestinal barrier. That which is absorbed is usually quickly eliminated by the liver [10]. Individuals with liver failure, however, are known to develop manganese overload and accumulate excess manganese in the brain [11]. In addition, those with iron deficiency absorb manganese much more avidly from the intestine and are therefore more susceptible to overexposure to the metal [12]. Inhalation of manganese circumvents the intestinal barrier and the liver. Thus absorption and deposition are an order of magnitude greater after inhalation rather than ingestion [8].

The insoluble manganese oxides, such as Mn_3O_4 are not easily cleared from the body after inhalation. In addition, clearance after low-dose exposure is much slower than after higher dose exposure. Animal studies have shown that the valence state of the manganese appears to greatly affect the metal's ability to enter the brain [8]. Mn3+ binds directly to transferrin and crosses the blood-brain barrier. Much of the manganese released from MMT combustion would be in the Mn3+ valence state. Once manganese enters the brain, whether through excessive oral absorption, inhalation, or liver failure, clearance is too slow to measure [13, 14].

In the body, manganese appears to seek out and bind to melanin. This pigment is found primarily in the hair, skin, eyes, and parts of the brain. In the brain, manganese is thought to catalyze oxidation of dopamine and may lead to membrane damage through free radical formation [15].

Newborns and fetuses avidly absorb manganese from the gastrointestinal tract and lack the ability to excrete the metal [16]. Manganese is also known to pass from the mother to the infant across the placenta and in breast milk [17]. In addition, a much higher percentage of manganese crosses the blood-brain barrier in infants than in adults [18]. These factors imply that infants and children may be particularly exposed, and also particularly susceptible to the toxic effects of manganese. It is important to remember, however, that the potential neurologic effects of this early exposure may not appear until many years later.

HEALTH EFFECTS OF MANGANESE

Manganese has long been known to be a health hazard to workers exposed through mining, in steel production, in the chemical industry, and in battery production. There appear to be three major end-organs for toxicity: the brain, the lungs, and the testes.

At high doses, such as those found in some workplaces, manganese causes a slow, degenerative neurologic condition which closely resembles Parkinson's disease [19]. This disease, known as manganism, begins with anorexia, fatigue, psychotic behavior, and clumsiness. It progresses slowly to an abnormal gait, excessive salivation, emotional lability, and slurred speech. The final stages include an expressionless, mask-like face, difficulty initiating movements, a shuffling gait, and tremors [20].

At lower levels of occupational exposure, subtle neurologic deficits, such as delayed reaction time, poor hand-eye coordination, memory deficits, and tremor have been reported. These have occurred in humans at doses as low as 150 $\mu g/m^3$ for an 8-hour TWA [21, 22]. Two studies even reported subtle neurological impairment in workers exposed to less than 100 $\mu g/m^3$ of respirable manganese [23-25]. The current OSHA PEL is 5000 $\mu g/m^3$ as a ceiling value, although the EPA has reported good evidence for neurotoxicity in adult workers at levels as low as 300 $\mu g/m^3$ [23]. The neurologic symptoms are known to worsen even after cessation of exposure [26].

Several studies suggest a subtle effect of manganese on behavioral characteristics and learning ability in children. Infants fed formula enriched with manganese had significantly higher hair manganese levels and had a significantly greater prevalence of hyperactivity and learning disability [27-29]. While infants are unquestionably a susceptible subpopulation, the elderly may also be at particular risk. With age, there is normally irreversible neuronal decline; this trend could be accelerated by low-level manganese exposure [30].

Inhalation of manganese is also toxic to the lungs, and apparently produces an inflammatory reaction as well as a suppression of pulmonary macrophage function. Exposed individuals have increased susceptibility to pneumonia and bronchitis [31]. Air levels below 20 $\mu g/m^3$ have been reported to increase the prevalence of respiratory symptoms (sputum, wheezing, sore throat) in schoolchildren [32].

Animal studies have shown that exposure to manganese during fetal development, at doses below those which cause other toxic effects, retards growth of the testes and reduces testosterone concentrations [33]. The research in male rodents is supported by similar findings in human workers. In one study, workers exposed at levels averaging one-fifth of the allowable workplace exposure limit had significantly fewer children during the period of exposure to the metal compared to similar unexposed workers [34]. However, a second study failed to confirm this finding [35]. There is, therefore, some evidence that manganese may adversely affect male fertility.

Lowest Reported Level for Health Effects in Humans

	Adults	Children
Subtle neurologic deficits:	100 $\mu g/m^3$	unknown
Respiratory symptoms:	1000 $\mu g/m^3$	7–87 $\mu g/m^3$
Reproductive effects:	90 $\mu g/m^3$	unknown

EXPOSURE ASSESSMENT

Some work has been done to quantify potential exposure levels resulting from use of MMT in gasoline. The percentage of total manganese eventually emitted via the tailpipe may range from 30 to 55 percent [36]. One study conducted by the Ethyl Corporation Research Laboratories concluded that at the highest levels of MMT use anticipated (0.125 g Mn/gal), airborne levels of Mn would increase 0.02–0.2 $\mu g/m^3$, with a median value of 0.05 $\mu g/m^3$. They also calculated that under the most unfavorable weather conditions, the estimated air concentration would be less than 3 $\mu g/m^3$ [9].

Two studies conducted in Canada, where MMT is used in gasoline, found airborne Mn levels in Montreal ranging from 0.01–0.42 $\mu\mu g/m^3$, with a mean of 0.036 $\mu g/m^3$ [37, 38]. More than 50 percent of the daily samples in the city of Montreal exceeded the urban background level of 0.04 $\mu g/m^3$ as established by the U.S. EPA [38].

Although these air levels do not seem high, it is important to remember that environmental levels of airborne Mn are normally extremely low. In the U.S., levels of airborne Mn fell from 0.07 $\mu g/m^3$ to 0.033 $\mu g/m^3$ between 1970 and 1982 [39]. While this decline has been attributed to control of point-source pollution from the burning of fossil fuels, some portion of this decrease may be due to the fact that MMT was added to leaded gasoline; the phase-out of leaded gasoline would be expected to be paralled by a decrease in manganese emissions from automobiles. The reintroduction of MMT, with its addition of an estimated 0.05 $\mu g/m^3$, would thus result in almost a tripling of levels of airborne Mn. Studies conducted in the U.S. during the short period of time in the 1970s when MMT was in use do in fact show significant changes in airborne Mn levels directly related to the introduction and then the removal of MMT from the U.S. fuel supply [40, 41]. MMT use would contribute approximately 2 million pounds of manganese annually to the environment [42].

Airborne manganese eventually settles to the ground where it contaminates the surface. One study conducted during the period of MMT use in the U.S. demonstrated an inverse relationship between Mn levels in street dust and proximity to a major traffic artery [43], suggesting that Mn concentrations on the ground may be significantly affected by increases in airborne levels. A more recent study in this country showed strong correlations between manganese concentrations in soil and distance from major roadways. The manganese concentrations were reported to be much higher than lead concentrations in roadside soil ever were [44]. Manganese may thus be expected to accumulate in the soil near roadways, which, over time. may lead to unacceptably high levels in the soil. These investigators estimated that between 0.01 and 0.8 grams of manganese per m^2 per year may reach the soil from MMT use [38]. One study of young children found that 1- to 4-year-olds normally ingest between 9 and 40 mg of soil per day, while children with pica may ingest as much as 5-8 grams of soil per day [45]. Thus, human exposures might be expected to occur via inhalation, ingestion of food grown in urban or roadside soil, and direct ingestion of soil in the case of small children.

Manganese accumulation in the environment may result in significantly increased intake of manganese in living organisms, with as yet unknown consequences. Manganese can accumulate in certain plants, a fact which has been demonstrated in an experimental study on oats and beans [46]. A study of pigeons in Canada has demonstrated that urban pigeons have higher manganese concentrations in their bodies than do rural pigeons, showing that increased body burdens may result from the increased ambient exposures [37]. It is unclear whether most of the manganese emitted from automobiles will remain on surface dust where it may intermittently become airborne again, whether it will be mostly taken up by plants, or whether it will enter groundwater in significant amounts. These questions illustrate areas requiring urgent investigation.

Environmental exposures to MMT itself should also be considered. Though MMT breaks down quickly on exposure to sunlight, it is highly volatile and much more toxic than manganese alone. Exposures could be expected to individuals pumping gasoline, particularly in gas stations which are not equipped with vapor recovery systems. Exposures are also likely to individuals who use gasoline as a cleaning solvent, and to those who inhale gasoline for recreational purposes.

WORKER EXPOSURES

Worker exposures may be a major problem in the cycle of MMT production and use. Manganese miners are at very high risk of neurological toxicity from chronic inhalation of manganese dusts. Neurological effects in miners have been reported to occur after only several months of exposure [13]. The production process of MMT is currently confined to one facility in South Carolina. There, elemental sodium is reacted with methylcyclopentadiene, and in turn with manganese chloride. Finally, carbon monoxide is added to form the end product [3]. Separation, purification, and preparation for shipping offer further opportunities for overexposure. Refinery workers are also at risk of overexposure to concentrated MMT. Although Ethyl Corporation does not report any worker overexposures, the Navy reported a series of overexposures of men handling C1-2™ additive. The men reported a strong metallic taste in the mouth (even after very slight skin exposure), nausea, headache, "giddiness," a "thick tongue," and profuse perspiration [2]. These symptoms suggest biological absorption and toxic effects, though the longer-term effects of such exposures are unknown. In addition, high-dose exposures to MMT would be likely to result in more severe toxicity and even death [47].

Once the compound is blended in gasoline, the exposures to MMT and to manganese would be much lower. However, the degree of risk to service station employees is unclear. One study of garage workers and taxi drivers in Montreal found significantly increased exposures to manganese on days at work compared to days off work, but the exposure levels (up to 2.07 $\mu g/m^3$) were well below the current occupational exposure limit of 5,000 $\mu g/m^3$ [48].

EFFECT ON CATALYTIC CONVERTERS

When the EPA originally prohibited the use of MMT in unleaded gasoline, it did so on the basis of concern that manganese oxides could foul catalytic converters, thereby reducing their effectiveness in controlling emissions. Early studies demonstrated increased levels of hydrocarbon emissions in cars fueled with MMT-treated gasoline. Other studies indicate that this increase may not be due to fouling of the converter, but may result from catalytic activity of manganese oxides upstream of the catalytic converter [49]. Ethyl performed a series of tests which showed no increase in hydrocarbon emissions from MMT-fueled cars. It was this series of studies which prompted EPA to conclude that MMT does not damage the catalytic converter.

However, the debate about the impact of MMT on automobile emissions continues. In *Ethyl Corporation v. E.P.A.* the American Automobile Manufacturers Association (AAMA) submitted a brief arguing that the evidence did indeed show that the "use of MMT as a fuel additive will cause or contribute to the failure of the emission control systems of some vehicles" [50]. The second generation of on-board diagnostics (OBD II) is now standard equipment on all new vehicles in the U.S. and Canada. This pollution control system appears to be disrupted by manganese build-up in the engine. The Ford Motor Company estimates that for every 50,000 miles traveled on MMT fuel, one-quarter pound of manganese is deposited in the catalytic converter [51]. According to the Motor Vehicle Manufacturers' Association (MVMA) of Canada, manganese is apparently encrusting the oxygen sensors, the noble metal in the catalyst, and spark plugs in cars throughout Canada, problems which can result in greatly increased hydrocarbon and carbon monoxide emissions, and in greatly shortened operating lifetimes of emissions control equipment [52]. In response to these findings, the AAMA has adopted specifications for gasoline recommending that consumers not use gasoline treated with additives containing manganese [53, 54], and the MVMA has called for legislation requiring the sale of MMT-free gasoline in Canada.

OCTANE ENHANCERS

MMT belongs to a group of gasoline additives known as "octane enhancers," whose addition to gasoline increases the octane number of the fuel. The higher the octane number, the greater the ability of the fuel to prevent engine knocking, which occurs when the fuel combusts prematurely in the engine. Engine knocking greatly increases mechanical and thermal stresses on the engine and is therefore undesirable. Almost all cars driven in the U.S. function efficiently using gasoline with an octane level of 87 (regular gas), yet most fuel companies advertise their higher octane gasolines as being better for the engine. Many Americans believe this advertising; and regularly purchase gasoline with an octane level higher than their cars need [55].

Unfortunately, this is false advertising, and the use of an octane level that is too high may result in the emission of unburned gasoline from the tailpipe.

There is currently a three-way dilemma in automobile engineering. The pressure to decrease automotive emissions led to the addition of catalytic converters which oxidize carbon monoxide and hydrocarbons. Unfortunately, these catalysts tend to somewhat reduce fuel economy. In order to improve fuel economy, new engine designs tend to have high compression ratios [56] to improve thermal efficiency [57]. An increased compression ratio increases the tendency of the engine to knock. This, in turn, requires a supply of higher-octane fuel and drives the development of octane-enhancers such as MMT [56].

There are two ways out of this dilemma. One involves engine re-design. There are several ways of re-designing the engine by controlling mixture movement in the combustion chamber or by supercharging the engine [59]. The so-called "lean-burn engine" shows some promise in solving some of these problems in the future but requires further development [60].

Alternatively, there are a range of methods available for increasing octane. Options include the addition of aromatic hydrocarbons, increased refining of the gasoline to increase the proportion of aromatics or branched-chain hydrocarbons, or addition of a variety of chemicals in relatively small amounts to serve as octane boosters. Butane, methanol, ethanol, ethyl tert-butyl ether (ETBE), and methyl tert-butyl ether (MTBE) may be added to gasoline to raise octane levels. The latter four chemicals boost octane and oxygenate fuel, thereby reducing carbon monoxide emissions by 10-20 percent from the engine. Ethanol, in particular, is renewable, and has no anticipated health effects. However, it is highly volatile, and must be blended into gasoline at a later stage outside the refinery [61]. In addition, there may be problems with vapor-lock and some decrease in driving performance from ethanol use [57]. Thus, addition of ethanol would be more expensive and less convenient [62]. MMT increases octane but does not oxygenate the fuel. It therefore would not decrease emissions, and may even increase hydrocarbon emissions. It is considerably less expensive than other methods of raising octane and therefore might be expected to dominate the market for octane enhancers if the market is allowed to operate freely [63].

CONCLUSION

MMT fuel additive is likely to contribute to an increase in ambient manganese concentrations, an increase in respirable manganese particulate, and some localized occupational and environmental overexposures to manganese and MMT. The health implications of widespread MMT use are not yet known. However, there are indications that certain highly exposed or susceptible subpopulations might be at increased risk for neurotoxicity, respiratory toxicity, and reproductive toxicity from manganese exposure. Such at-risk populations include infants, the elderly, those with iron deficiency or liver disease, and those who are highly exposed to gasoline or engine exhaust through work or a hobby. In addition, there are very significant health hazards to workers in manganese mining, MMT production, and petroleum blending from high-dose exposures. There is considerable evidence that MMT use results in manganese deposits in the pollution-control systems of newer-generation cars, and that ultimately such deposits are likely to lead to greater automotive emissions, and premature failures of catalytic converters. Alternatives to MMT are readily available, and newer generations of engine technology may not require octane-enhanced fuel.

Fifteen oil companies which together produce more than 70 percent of U.S. gasoline currently claim that they are voluntarily withholding MMT from their product [64]. Unfortunately, only four of the companies have promised to notify the Environmental Defense Fund if they do decide to start using MMT [65]. Thus, there is still reason for concern and a need for more decisive action. The lower cost of MMT, and the fact that it is readily available and legal, may drive refiners to use it despite concerns over health and engine-related effects. The Canadian Minister for the Environment has introduced legislation to ban MMT in Canada [66]. Unfortunately, however, EPA approval opens the door for use of this additive throughout the world, with attendant risks to millions of people. The dispersal of an incompletely tested substance throughout the environment is unfortunately common, and has proved to be a mistake on numerous occasions; it remains unethical and irresponsible and should prompt reconsideration of our chemical regulatory processes.

SUGGESTIONS FOR ACTION

RESEARCH

- Investigation of potential biomarkers of manganese exposure to allow for improved exposure assessment in both occupational and environmental epidemiologic studies.
- Further research on the health effects of low-dose exposure to manganese, particularly on thresholds for neurotoxicity, respiratory toxicity, and reproductive toxicity.
- Research on MMT-exposed workers to look for health effects particular to MMT.
- Further investigation of the distribution and fate of manganese in the environment and its effects on ecosystems.

REGULATION

- The OSHA PEL and ACGIH TLV should both be lowered from 5000 to 150 μg/m^3 as an 8-hour TWA.
- The EPA should initiate action under section 211(c) of the CAA to withdraw licensing of MMT.
- Congress should clarify section 211(f)(4) of the CAA to allow EPA to consider public health effects of gasoline additives prior to licensing.

ACTIVISM

- Pressure Ethyl Corporation to withdraw MMT from the market.
- Pressure petroleum refiners to label gasoline MMT-free at the pump.
- Develop and implement a rapid test for the presence of manganese in gasoline to allow independent verification of the absence of MMT by local public interest organizations.
- Advocate a precautionary approach to introduction of new chemicals which would require adequate testing prior to marketing and dispersal.

ACKNOWLEDGMENTS

Supported by the Occupational Safety and Health Educational Resource Center, CDC/NIOSH grant award T42/CCTl 10421 at the Harvard School of Public Health and by the Occupational Physicians Scholarship Fund. The authors would like to thank Karen FIorini and Lawrie Mott for their comments and advice.

REFERENCES

1. 51F.3d 1053 (D.C. Cir. 1995)
2. Editorial Staff, Ever Hear of Methyl-Cyclopentadienyl Manganese-Tricarbonyl? *Mech* (publication of the Naval Safety Center), Summer 1970.
3. Ethyl Corporation, *MMT Information for the U.S. EPA Chemical Hazard Information Profile,* August 1983, EPA/OTS; Doc #FYI-OTS-0883-0253.
4. 42 U.S.C.A. § 7545(f)(4)
5. 51F.3d 1053 (D.C. Cir. 1995)
6. 67F.3d 941 (D.C. Cir. 1995)
7. W. Moore et al., Metabolic Aspects of Methylcyclopentadienyl Manganese Tricarbonyl in Rats, *Environmental Research, 8,* pp. 171-177, 1974.
8. G. Wood and M. Egyed, *Risk Assessment for the Combustion Products of Methylcyclopentadienyl Manganese Tricarbonyl (MMT) in Gasoline,* Environmental Health Directorate, Health Canada, December 6, 1994.
9. G. L. Ter Haar et al., Methylcydopentadienyl Manganese Tricarbonyl as an Antiknock: Composition and Fate of Manganese Exhaust Products, *Journal of the Air Pollution Control Association, 25,* pp. 858-860, 1975.

10. G. L. Rehnberg et al., Chronic Ingestion of Mn304 by Rats: Tissue Accumulation and Distribution of Manganese in Two Generations, *Journal of Toxicology and Environmental Health, 9,* pp. 175-188, 1982.
11. R. A. Hauser et al., Blood Manganese Correlates with Brain Magnetic Resonance Imaging Changes in Patients with Liver Disease, *Canadian Journal of Neurological Science, 23,* pp. 95-98, 1996.
12. S. V. Chandra and S. K. Tandon, Enhanced Manganese Toxicity in Iron-Deficient Rats, *Environmental Physiology and Biochemistry, 3,* pp. 230-235, 1973.
13. J. Donaldson, The Physiopathologic Significance of Manganese in Brain: Its Relation to Schizophrenia and Neurodegenerative Disorders, *Neurotoxicology, 8,* pp. 451-462, 1987.
14. G. C. Cotzias et al., Chronic Manganese Poisoning: Clearance of Tissue Manganese Concentrations with Persistence of the Neurological Picture, *Neurology, 18,* pp. 376-382, 1968.
15. D. Mergler, Manganese: The Controversial Metal. At What Levels Can Deleterious Effects Occur? *Canadian Journal of Neurological Science, 23,* pp. 93-94, 1996.
16. G. C. Cotzias et al., Interactions Between Manganese and Brain Dopamine, *Medical Clinics of North America, 60,* pp. 729-738, 1976.
17. S. Tabacova, Maternal Exposure to Environmental Chemicals, *Neurotoxicology, 7,* pp. 421-440, 1986.
18. W. S. Webster and A. A. Valois, Reproductive Toxicology of Manganese in Rodents, *Neurotoxicology, 8,* p. 437, 1987.
19. A. Barbeau, Manganese and Extrapyramidal Disorders, *Neurotoxicology, 5,* pp. 13-36, 1984.
20. I. Mena et al., Chronic Manganese Poisoning: Clinical Picture and Manganese Turnover, *Neurology, 17,* pp. 128-136, 1967.
21. A. lregren, Using Psychological Tests for the Early Detection of Neurotoxic Effects of Low Level Manganese Exposure, *Neurotoxiciology, 15,* pp. 671-677, 1994.
22. H. Roels et al., Epidemiological Survey Among Workers Exposed to Manganese: Effects on Lung, Central Nervous System, and Some Biological Indices, *American Journal of Industrial Medicine, 11,* pp. 307-327, 1987.
23. L. S. Erdrich and J. P. Stara, *Health Assessment Document for Manganese: Project Summary,* U.S. EPA EP 1.89/2:600/S8-83/013F: July 1987.
24. D. Mergler et al., Nervous System Dysfunction among Workers with Long-Term Exposure to Manganese, *Environmental Research, 64,* pp. 151-180, 1994.
25. A. Iregren, Psychological Test Performance in Foundry Workers Exposed to Low Levels of Manganese, *Neurotoxicology and Teratology, 12,* pp. 673-677, 1990.
26. C. C. Huang et al., Progression After Chronic Manganese Exposure, *Neurology, 43,* pp. 479-1483, 1993.
27. R. O. Pihl and M. Parkes, Hair Element Content in Learning Disabled Children, *Science, 198,* pp. 204-206, 1977.
28. P. J. Collipp, S. Y. Chen, and C. S. Maitinsky, Manganese in Infant Formulas and Learning Disability, *Annals of Nutrition and Metabolism, 27,* pp. 488-494, 1983.
29. P. J. Barlow and M. Kapel, *Hair Metal Analysis and its Significance to Certain Disease Conditions,* 2nd Annual Trace Minerals Health Seminar, Boston, 1979.
30. E. K. Silbergeld, Current Status of Neurotoxicology, Basic and Applied, *TINS,* pp. 291-294, 1982.

31. P. J. Abbott, Methylcyclopentadienyl Manganese Tricarbonyl (MMT) in Petrol: The Toxicologic Issues, *Science of the Total Environment, 67,* pp. 247-255, 1987.
32. K. Nogawa et al., Epidemiological Studies on Disturbance of Respiratory System Caused by Manganese Air Pollution. Report 1: Effects on Respiratory System of Junior High School Students, *Japanese Journal of Public Health, 20,* pp. 315-326, 1973.
33. L. E. Gray and J. W. Laskey, Multivariate Analysis of the Effects of Manganese on the Reproductive Physiology and Behavior of the Male House Mouse, *Journal of Toxicology and Environmental Health, 6,* pp. 861-867, 1980.
34. R. Lauwerys et al., Fertility of Male Workers Exposed to Mercury Vapor or to Manganese Dust: A Questionnaire Study, *American Journal of Industrial Medicine, 7,* pp. 171-176, 1985.
35. J-P. Gennart, J-P. Buchet, H. Roels, et al., Fertility of Male Workers Exposed to Cadmium, Lead, or Manganese, *American Journal of Epidemiology, 135,* pp. 1208-1219, 1992.
36. S. Loranger, J. Zayed, and E. Forget, Manganese Contamination in Montreal in Relation with Traffic Density, *Water, Air, and Soil Pollution, 74,* pp. 385-396, 1994.
37. S. Loranger et al., The Pigeon (Columba livia) as a Monitor for Manganese Contamination from Motor Vehicles, *Archives of Environmental Contamination and Toxicology, 27,* pp. 311-317, 1994.
38. S. Loranger, J. Zayed, and E. Forget, Manganese Contamination in Montreal in Relation with Traffic Density, *Water, Air, and Soil Pollution, 74,* pp. 385-396, 1994.
39. D. R. Lynam et al., Environmental Assessment of MMT Fuel Additive, *Science of the Total Environment, 93,* pp. 107-114, 1990.
40. W. R. Pierson et al., Methylcyclopentadienyl Manganese Tricarbonyl: Effect on Manganese Emissions from Vehicles on the Road, *Journal of the Air Pollution Control Association, 28,* pp. 692-693, 1978.
41. P. J. Lioy, R. P. Mallon, and T. J. Kneip, Long-Term Trends in Total Suspended Particulates, Vanadium, Manganese, and Lead at Near Street Level and Elevated Sites in New York City, *Journal of the Air Pollution Control Association, 30,* pp. 153-156, 1980.
42. Ethyl Corporation Document, *MMT and Public Health: Separating Fact from Fiction,* p. 10, May 14, 1996.
43. M. M. Joselow et al., Manganese Pollution in the City Environment and its Relationship to Traffic Density, *American Journal of Public Health, 68,* pp. 557-560, 1978.
44. C. Lytle et al., Manganese Accumulation Along Utah Roadways: A Possible Indication of Motor Vehicle Exhaust Pollution, *Science of the Total Environment, 162,* pp. 105-110, 1995.
45. E. J. Calabrese et al., How Much Soil Do Young Children Ingest: An Epidemiologic Study, *Regulatory Toxicology and Pharmacology, 10,* pp. 123-137, 1989.
46. N. Brault et al., Bioaccumulation of Manganese by Plants: Influence of Methylcyclopentadienyl Manganese Tricarbonyl as a Gasoline Additive, *Science of the Total Environment, 153,* pp. 77-84, 1994.
47. D. N. Cox et al., Comparison of the Toxicity of Methylcydopentadienyl Manganese Tricarbonyl with that of its Two Major Metabolites, *Toxicology Letters, 39,* pp. 1-5, 1987.

48. J. Zayed et al., Occupational and Environmental Exposure of Garage Workers and Taxi Drivers to Airborne Manganese Arising from the Use of Methylcyclopentadienyl Manganese Tricarbonyl in Unleaded Gasoline, *American Industrial Hygiene Association Journal, 55,* pp. 53-58, 1994.
49. K. Otto and R. J. Sulak, Effects of Mn Deposits from MMT on Automotive Catalysts in the Absence and Presence of Other Fuel Additives, *Environmental Science and Technology, 12,* pp. 181-184, 1978.
50. 51 F.3d 1053 (D.C. Cir. 1995)
51. American Automobile Association, MMT in Gasoline: Profits and Problems?, *Autogram #20,* April 1996.
52. Motor Vehicle Manufacturers' Association, *The Impact of Manganese-Based Fuel Additives on Vehicle Emission Control Technology in Canada,* Toronto, Ontario, October 1995.
53. Environmental Defense Fund, *Producers of Over 70% of US Gasoline Not Using MMT,* New York, March 18, 1996.
54. AAMA, *AAMA Gasoline Specification,* February 5, 1996.
55. B. Mohl, Few Cars Need Premium Gasoline: Ads Misleading, Analysts Warn, *Boston Globe,* p. 13, October 2, 1995.
56. The ratio of the volume of air and fuel when the piston is at the lower end of the cylinder to the volume when the mixture is compressed and the piston is in its topmost position.
57. The ratio of power obtained by the engine to the maximum power available in the fuel.
58. Royal Dutch/Shell Companies, *The Petroleum Handbook* (6th Edition), Elsevier, 1983.
59. Compressing the air and fuel mixture before introduction into the cylinders.
60. U. Seiffert and P. Walzer, *Automobile Technology of the Future,* Society of Automotive Engineers Inc., Warrendale, Pennsylvania, 1991.
61. GAO, *Gasoline Marketing: Uncertainties Surround RFG as a Motor Fuel,* GA 1.13: RCED-90-153, June 1990.
62. Health Effects Institute, *The Potential Health Effects of Oxygenates Added to Gasoline: A Review of the Current Literature,* HEI, Boston, February 1996.
63. Anonymous, Which Antiknock Will It Be—Aromatics or MMT?, *Chemtech,* pp. 484-487, August 1978.
64. Amoco, Anchor, ARCO, BF Chevron, Conoco, Exxon, Hess, Marathon Oil, Mobil, Pennzoil, Phillips, Shell, Sun, and Texaco.
65. Amoco, Anchor, ARCO, and Sunoco.
66. Environment Canada News Release, *Government Reintroduces MMT Bill C-94,* April 18, 1996.

Note: Originally published in *New Solutions* (Winter) pp. 17-25, 1999.

CHAPTER 9

Describing Community Health Risks: Can Epidemiology Be Improved?

David Kriebel

Environmental threats are one of the highest concerns of communities seeking to improve public health. University researchers seek to assist communities with these concerns by providing technical expertise. But these community-science collaborations are often frustrating to both parties. Communities often find that the scientific studies have not "proven" what they believe the problem to be, while scientists often feel at least pestered, if not manipulated, by the "dirty" politics of the community. This paper investigates the proposition that environmental scientists must learn to combine their quantitative studies with qualitative views on the community and its health, if their studies are to be useful and indeed even fully accurate. The goal of the presentation is to explore changes in the methods of one particular field, environmental epidemiology, that might make it more relevant to current environmental debates. After a brief history of the field, limitations in the current practice of epidemiology are presented. These can be grouped under two headings: an over-emphasis on quantitative results, and an under-estimation of uncertainties in study results. Some suggestions are given for changes that might address these problems, and make epidemiology more useful for citizens and policy-makers. These are:

1. nest quantitative results in qualitative descriptions of study populations;
2. fit two different kinds of statistical models—one very simple, the other quite complex;
3. conduct sensitivity analyses; and
4. use precautionary guidelines for causal inference.

It is concluded that environmental epidemiology serves a particular societal function; which is to inform citizens and policy makers about environmental hazards that, perhaps, should be the subject of legal and economic interventions. For this reason, methods could be changed to improve the fit of the science to this function. There remains, however, much to be done to revise and refine this perspective, and to make the suggestions sufficiently specific that they can be put into practice.

Environmental threats are one of the highest concerns to communities seeking to improve public health. University researchers seek to assist communities with these concerns in a number of different ways. Sometimes academics focus on information: making technical content accessible to the public, so that they may make informed decisions. A second approach addresses community empowerment, so that voices of concern are heard beyond the boundaries of the neighborhood, and community political power is enhanced. Another important contribution is research seeking to quantify the magnitude of risks, and perhaps quantify benefits as well. Indeed, community groups that have targeted environmental threats often ask for technical assistance from researchers in such fields as air pollution monitoring, epidemiology, and toxicology. But these community-science collaborations are often ultimately frustrating to both parties. Communities often find that the scientific studies have not "proven" what they believe the problem to be, while scientists often feel at least pestered, if not manipulated, by the "dirty" politics of the community.

This chapter begins from the following proposition: that environmental scientists must learn to combine their quantitative studies with qualitative views on the community and its health, if their studies are to be useful and indeed even fully accurate. The goal of the presentation is to explore changes in the methods of scientific research that might make it more relevant to current environmental debates. So as to be as concrete as possible about the limitations of current methods, and the changes that might be made, the paper focuses on environmental epidemiology, one of the important branches of environmental science. The presentation proceeds as follows: a brief history is provided to show how epidemiology has evolved into the powerful quantitative tool that it now is, then a critique of the limits of the field is presented. Finally, an outline is developed, to suggest how environmental epidemiology might be changed to make it more useful to environmental policy debates.

A BRIEF HISTORY OF EPIDEMIOLOGY

Epidemiology has a long and proud history as the principal quantitative science of public health [1, 2]. Prior to the last half of the twentieth century, the primary contributions were in the study of mortality, and in the transmission of infectious diseases. Perhaps the most important understanding that the early epidemiologists gained was the realization that diseases could profitably be studied in populations, rather than simply in the afflicted individuals. When viewed from the distance that the calculation of population rates permitted, diseases and mortality were seen to behave quite rationally, despite the terrible randomness of an epidemic of plague, or cholera, or the tragic risks of childbirth. In the most famous early example, the

London physician John Snow identified patterns of cholera mortality in London in 1854 which enabled him to see a link to contaminated water which so many thousands of other Londoners could not see, despite being surrounded by the casualties of horribly polluted water [3].

Early epidemiology (essentially prior to the Second World War) was a largely descriptive and synthetic discipline, practiced by physicians [4]. They used a wide variety of different sources of evidence to formulate and then test hypotheses about the distribution and determinants of diseases. They calculated rates of disease in populations, but also carefully studied individual patients. They observed air, water, and food, climatologic data, and any other relevant information. They did not, in general, concern themselves with quantitative exposure-risk relationships, which have become nearly the sole focus of the field today.

Modern epidemiology arose out of these historical roots as a more quantitative, and more formally constructed set of methods for detecting and quantifying exposure-disease relationships. One of the key developments that changed epidemiology was the tremendous success of randomized clinical trials in the early post-World War II period. Trials like those that demonstrated the effectiveness and safety of the polio vaccine made a powerful impression on epidemiologists. A randomized clinical trial is an experiment, in which treatment and placebo groups are randomly chosen. If there are differences in outcome between these two groups, they are highly likely to be due to the treatment, because randomization has insured that the two groups are probably quite similar on all other factors that might explain a difference between the two groups. The problem of an unintended difference between two groups that are being compared to look for the effect of some treatment or exposure is called confounding in epidemiology, and randomization is a very effective way to control confounding. But most epidemiologic studies cannot use randomization—we cannot randomly assign smoking or seat-belt wearing, for example. So, the challenge to the epidemiologists of the 1960s and '70s was to figure out how to control confounding in observational studies— those in which the exposures are chosen by social forces and individual choices and not by the researcher. Methods for the statistical control of confounding were developed and codified in the 1970s, and this development more than any other led to the belief that observational epidemiologic studies could approach (even if never quite achieve) the inferential power of the randomized trial.

The formal comparison of an exposed and an unexposed population, again inspired by the treatment and placebo groups of the clinical trial, led also to the standardization of the concept of the measure of association. Also called a measure of effect, and most frequently represented by a "relative risk," this was seen as the key statistic capturing the essential information about the magnitude of the risk to be attributed to a particular exposure. For example, the risk of lung cancer faced by a lifelong cigarette smoker is estimated to be about 20 times the risk for a nonsmoker. That is, the relative risk for smoking is 20.

Statistical control of confounding and quantification of measures of association are largely accomplished through the fitting of statistical models to large masses of health and exposure data. This process was greatly facilitated, and in practice made technically feasible, starting in the 1970s by the widespread introduction of computers. Without substantial, inexpensive computing power, epidemiology would likely have remained a much less quantitative and more descriptive discipline. Thus, the theoretical advances in epidemiology occurred nearly simultaneously with the technological revolution that made implementation of the new statistical methods and modeling techniques possible.

From this brief overview, one can see that while the main themes of epidemiology date from the 18th and 19th centuries, it is really only in the last 25 years or so that the field has become focused so singularly on the quantification of statistical measures linking exposure to disease. That this key change in the field is so recent may provide some encouragement for those who would contemplate trying to influence current practice of environmental epidemiology.

SOME LIMITATIONS OF ENVIRONMENTAL EPIDEMIOLOGY

An Over-Emphasis on Quantitative Results

With these developments in epidemiologic methods came a greater emphasis on generalizability of study results. That is, the ability to control confounding, to fit complex computer models, and to summarize the results in a few standard statistics made it much easier to quantitatively compare one study to another. A table of relative risks for a series of studies is a highly efficient way to summarize a series of studies. But are these studies all comparable? Is there information about the particulars of the populations studied that are relevant to the interpretation of the results that are lost in building the table? There are many ways that differences in a study's setting and population may affect its quantitative results. One example is that environmental exposures to a hazardous chemical are rarely, if ever, fully comparable from one study to another. A series of studies of say fine particles in urban air and mortality may all use the same measuring instruments, but the particular mix of industries and meteorologic conditions in each city makes it difficult to be confident that the results are fully comparable from one city to the next. If the results of these studies are compared using a simple summary statistic like a relative risk for each city, then we are making the assumption that the only important source of differences between these numbers is sampling variability around some "true" relative risk. How reasonable is this assumption? It's difficult to say, because we can never know the "true" risk, and so cannot see how biased our estimates of it really are.

The tendency to focus on relative risks, measures of association, leads to loss of potentially important qualitative, descriptive information about each population under study. There is no place in modern epidemiology for all of the "messy" interpretation of the particulars of a population and its unique situation. For example, we recently completed a study of the respiratory health of workers in an automotive factory in Indianapolis [5]. The hypothesis was that exposure to metal-working fluids (mists of these chemicals are generated in the air around machines that cut metal) would adversely affect the lungs of workers. The factory chosen for study turned out to have much lower levels of exposure to metal-working fluids than had been expected. Perhaps not coincidentally, there were only minor effects on lung function detectable among the exposed workers. The study was thus interpreted as largely "negative." And yet, there was evidence of effects on respiratory health. For example, it appeared likely that workers who had been in the exposed areas had moved to other jobs where they were less exposed to these chemicals. There was also evidence that people were sicker in the unexposed areas. It appeared likely that workers had suffered from the exposures, and then transferred to cleaner parts of the plant. But with standard epidemiologic methods, we could not quantify this, and thus there was no way to effectively incorporate it into the study findings.

Under-Estimation of Uncertainty

There are many sources of uncertainty in epidemiologic studies. These include:

- errors in exposure estimation;
- incomplete control of confounding;
- biases;
- model form mis-specification; and
- sampling variability in the dependent variable.

Only the last of these is formally evaluated in a typical epidemiologic study. The first, errors in exposure estimation, is a major problem in environmental epidemiology, but has only recently drawn formal recognition, albeit in a very limited number of studies. Incomplete control of confounding refers to the fact that the statistical methods to control confounding that were mentioned earlier can only work if a potential confounding factor is recognized, and information on it is gathered. If, alternatively, there is an unknown risk factor that differs between the two groups being compared, its confounding effect cannot be controlled. Only randomization has the ability to control for the confounding influence of factors that are not measured or even recognized. This problem is well-described in every textbook, but in conventional practice, we often behave as if unmeasured confounders are unlikely to be a serious problem, when there is no way to really know this. Thus, there is uncertainty that derives from what we don't know—and this is by definition very difficult to quantify. Uncertainty deriving from possible bias in study design and execution is quite similar in nature—there is rarely any way to objectively know whether a bias exists, nor how strong its effect on the results might be. It is standard practice in reporting the results of an epidemiologic study to say at the end of the chapter that the results are only valid if there is no bias and no uncontrolled confounding. But since this is an unverifiable assertion, it is often essentially ignored, in the evaluation of the evidence. But this habit gives the non-scientist the impression that we are more confident than we really are.

There has been very little discussion in epidemiology of the uncertainties that may result from model form mis-specification. In quantitative epidemiologic studies, there is always at least one model imposed between the data and us. There is some (mathematical or statistical) model which we use to summarize the data and which allows us to make inferences—like the estimate of the strength of the association described previously. The question is: what is the right form of that model? And what happens if we choose the wrong form? A recent simulation study investigated this question [6]. It concluded that at best we can consider the results from a mis-specified model as an accurate description of the population under study. But the results from a mis-specified model, extrapolated to another population, may give quite misleading results. Unfortunately, very little is known about which models are best for particular settings, and so the choice is often based on convenience.

The final source of uncertainty, sampling variability in the dependent variable is the only source that is routinely estimated. All epidemiologic studies now calculate either a *p*-value or a confidence interval that provides an estimate of the degree to which chance is likely to explain the observed measure of association. The familiar "statistical significance" of a result is based on these calculations. The problem is that sampling variability—the amount that the result would likely vary from one hypothetical repetition of the study to another, is probably a small part of the true uncertainty in the study result, and yet we treat it as if it is *the* measure of the uncertainty.

One should not be surprised to learn that researchers are reluctant to discuss additional uncertainties in their results. They fear that their work will be further weakened, and their results even less useful. What is the solution? I propose that this is where the qualitative, descriptive methods need to be incorporated into environmental research; a full presentation of what is known about a particular environmental risk would include both the quantitative results, with all their uncertainty, plus additional information that is currently lost in the exclusive focus on the quantitative.

A MODEST PROPOSAL FOR CHANGE

What can be done to make epidemiology more useful for environmental policy debates? Here are some suggestions.

Nest the Quantitative in the Qualitative

There is certainly a role for the quantitative estimates of the degree of evidence for an association between a hazard and a disease. But we need to find ways to complement these measures—inevitably fraught with much uncertainty—with qualitative information that fills out the picture of the population and the environment that have been studied. It may be useful to think about this as "nesting." The quantitative measures are narrow in their view, limited in their generalizability, but highly economical in their presentation. Perhaps these should be wrapped in richer descriptions of the particulars of the study; details which will help the citizen or decision-maker evaluate the relevance of the quantitative results for the specific decision that must be made.

The other suggestions are directed toward improving the quantitative methods, and adapting them better to the task of informing environmental policy.

Fit Two Different Kinds of Models

First, it may be useful to fit very simple models that make as few assumptions as possible, but which are also less easily interpretable. For example, one might simply compare broad groups with obviously different exposure characteristics—garment workers and non-garment workers, while recognizing fully that this comparison captures far more than a difference in, for example the postures in which they work. Second, one might conduct highly detailed studies (necessarily of a small number of workers) to quantify the forces and angles of posture, honing in on a narrowly specified hypothesis about these "exposures" and risk of musculoskeletal disease. This model might be further specified with a mathematical model designed to use these exposure data, but also to capture the essential behavior of a hypothesized biologic mechanism by which these exposures may lead to a particular disease. The first model is useful because its "face value" is high; there are few assumptions, and so fewer places where it can be misleading. But it is also of limited value for the policy-maker or engineer who would, for example, try to redesign a machine to reduce injuries. The second model has many more ways in which it may be wrong, but is also much more useful for prevention (if it can be assumed to be correct).

Conduct Sensitivity Analyses

How much do one's results change when one changes the assumptions underlying the methods used? A standardized approach to sensitivity analysis that was included in every study would help the user evaluate the strength of the evidence provided. There are several impediments to this. First, there are no accepted guidelines for what kinds of analyses to do, so one is always suspicious that the only reported analyses were the ones that supported the researcher's own biases.

Second, scientific journals want articles to be short, and concise. Sensitivity analyses add length, and ambiguity to a study.

Use Precautionary Guidelines for Causal Inference

Finally, what about causal inference? The full weight of evidence should be considered when evaluating the evidence for causality. This means both qualitative and quantitative evidence. Epidemiologists use a fairly standard set of criteria in evaluating causality that are called the Hill criteria [7]. These criteria seem to be quite useful, but in recent years they have often been interpreted quite narrowly. A more precautionary interpretation is possible, and is outlined here.

1. *Temporal sequence.* Hill said that one should always evaluate epidemiologic evidence to insure that one could conclude unambiguously that the exposure preceded the effect, and not the other way around. This common-sensical consideration is clearly reasonable, and requires no modification.

2. *Strength of association.* Strong associations are considered more likely to be causal, and less likely to be due to chance or bias than weak ones, and this seems like a reasonable and appropriate consideration. Traditionally, strength is assessed by the measure of association alone. But, as discussed above, this may not be sufficient for evaluating strength—there may be relevant qualitative information that should be considered.

3. *Consistency of results across studies.* Results which are found repeatedly in many studies are considered more likely to be causal. This seems quite reasonable, as observational studies are fraught with many potential biases. But this criterion should be tempered by considerations of the consequences of not acting on a single, or a limited number, of studies. Because there is no such thing as scientific certainty, one must always weigh the costs of failing to act in the face of uncertainty.

4. *Biologic plausibility.* Hill said that epidemiologic evidence of an exposure-disease association was more likely to indicate a true causal link if there was a plausible mechanism by which this exposure could be understood to act. Usually this means that the effect has been seen in laboratory animals. Again, this seems quite reasonable, but this criterion has sometimes been too narrowly defined. Because there is so much that is still unknown about how most environmental chemicals react in biologic systems, it seems imprudent to require evidence of the same disease in animals that is observed in humans. Rather, evidence of toxicity or biologic activity, even if it would appear to be acting by a different biologic pathway, may be sufficient to provide biologic plausibility.

5. *Exposure response.* This criterion says that if there is evidence that increasing amounts of exposure lead to increased risk of disease, then the observed association is more likely to be causal. This is a reasonable assumption, but one that has been too rigidly interpreted. Because of the many sources of error in

exposure estimation, one should not necessarily expect to see simple exposure-response relationships in epidemiologic data. Thus, absence of a clear exposure-response trend should not be taken as strong evidence against causality.

CONCLUSION

Environmental sciences, including environmental epidemiology, serve a particular societal function. They do not exist solely to increase human understanding of the world. They may serve a variety of functions, but one of the most important is to inform citizens and policy-makers about environmental hazards that, perhaps, should be the subject of legal and economic interventions. This paper has presented a perspective on the "fit" of environmental epidemiology to this function. It is concluded that methods could be changed to improve the fit, and some suggestions for changes have been presented. There remains, however, much to be done to revise and refine this perspective, and to make the suggestions sufficiently specific that they can be put into practice.

REFERENCES

1. N. Krieger, Epidemiology and the Web of Causation: Has Anyone Seen the Spider?, *Social Science in Medicine, 39*, pp. 887-903, 1994.
2. J. P. Koplan, S. B. Thacker, and N. A. Lezin, Epidemiology in the 21st Century: Calculation, Communication, and Intervention, *American Journal of Public Health, 89*, pp. 1153-1155, 1999.
3. W. H. Frost, *Snow on Cholera*, Hafner, New York, 1965.
4. R. Stallones, To Advance Epidemiology, *Annual Review of Public Health, 1*, pp. 69-82, 1980.
5. D. Kriebel, S. R. Sama, S. Woskie, D. C. Christiani, E. A. Eisen, S. K. Hammond, D. K. Milton, M. Smith, and M. A. Virji, A Field Investigation of the Acute Respiratory Effects of Metal Working Fluids, I. Effects of Aerosol Exposures, *American Journal of Industrial Medicine, 31*, pp. 756-766, 1997.
6. G. Maldonado and S. Greenland, Impact of Model-Form Selection on the Accuracy of Rate Estimation, *Epidemiology, 7*, pp. 46-54, 1996.
7. C. Hennekens and J. Buring, *Epidemiology in Medicine*, Little Brown, Boston, 1987.

Note: Originally published in *New Solutions, 11*(1), pp. 13-21, 2001.

CHAPTER 10

Have Risks Associated with the Presence of Synthetic Organic Contaminants in Land-Applied Sewage Sludges Been Adequately Assessed?

Robert C. Hale and Mark J. La Guardia

Land application has become the dominant means for sewage sludge disposal in the United States. In 1993, the EPA concluded that synthetic organics therein posed an insignificant risk, based on the results of the 1988 National Sewage Sludge Survey, the view that most persistent organics are no longer in commerce and that industrial pretreatment further reduces their release to municipal treatment plants. However, we detected high concentrations of several problematic compounds in biosolids that were overlooked in the risk assessment, including persistent, bioaccumulative polybrominated diphenyl ethers and estrogenic nonylphenols. These results and other findings call into question the assessment's certainty.

Application of sewage sludge on agricultural, public, and reclaimed lands has become the dominant means for its disposal in the United States. While wastewater contaminants concentrate in sludge, a 1993 U.S. EPA risk assessment concluded that land application of sludge-associated synthetic organics does not pose significant risks and thus need not be regulated. EPA based this on a 1988 analytical survey that failed to detect significant burdens of targeted organics in sludge, the view that persistent organics of concern have been removed from commerce and that industrial pretreatment limits release of problematic organics to treatment plants. However, we detected high concentrations of polybrominated diphenyl ethers and nonylphenols, recently recognized as contaminants of concern, in sludges from several U.S. states. These pollutants were not considered in the EPA risk assessment. Potential additive effects of the myriad of contaminants in sludge were also not evaluated. These results call into question the certainty of the assessment conclusions.

Effective treatment of wastewaters is an onerous task due to the tremendous volumes generated and the diversity of chemicals contained therein. It has been estimated that as many as 100,000 chemicals are in commerce and thus can potentially enter wastewater [1]. In addition, about a thousand new compounds are introduced annually, making static environmental regulations untenable. Degradative processes are essential elements of the treatment process, but often are incomplete, generating additional intermediates of toxicological concern. While the goal of the treatment process is to produce an aqueous effluent free of short- or long-term environmental or health effects, an additional product is a solid residue. This sludge is formed by flocculation of dissolved and suspended materials originally present in the wastewater. Hydrophobic chemicals preferentially sorb to the particulates and associated organic matter and are then removed from the water column by settling. Thus, sludge is enriched not only in nutrients and organic carbon, but also pollutants.

Approximately 6.9 million dry tons of sewage sludge were generated in the United States in 1998 [2]. The amount is expected to increase as wastewater treatment improves and the population served grows. Disposal of this material is a significant issue. Sludge may be incinerated. However, facilities required are expensive and must be properly engineered and maintained to minimize production of by-products, such as chlorinated dioxins. Landfilling of sludge has been portrayed as a waste of disposal capacity and valuable real estate. While the number of landfills has decreased from 8,000 to 2,300 from 1988 to 1999, total capacity has remained relatively constant [3]. If all U.S. sludge was landfilled, it would constitute only about 5 percent of the total dry weight of solid waste disposed of in that manner. A third option, ocean disposal, fell into disfavor largely due to incidents involving debris on northeastern U.S. shorelines and related beach closings. Sanctioned ocean disposal of sludge ended in 1992 as a result of the Ocean Dumping Act. Factors favoring land application of sewage sludge include reutilization of the nutrients contained therein and enhancement of soil porosity and organic content. Land application is often the most economic means of sludge disposal compared to the alternatives, at least for the generator. As a result it has become the dominant disposal mechanism in the United States, encompassing more than half of the wastewater treatment plant (WWTP) sludge produced. After generation at the WWTP, sewage sludge typically is further processed or "stabilized" to reduce its water and pathogen content. The term "biosolids" has been coined for this product. Biosolids are subsequently given away or sold at low cost to farmers, the general public or others.

However, land application of sludge may present health risks, due to the presence of pathogens, toxic metals, and organic pollutants. Pathogens may remain viable in sludge to varying degrees as a function of the stabilization technique employed. Toxic metals do not breakdown in the treatment process or in the environment. As a consequence, they can build up in soil upon repeated application and thus are regulated by the EPA Part 503 regulations [4]. Nonetheless, limitations on metal burdens in U.S. biosolids are among the most lenient in the world [5]. No U.S. regulations exist limiting organic pollutant burdens. The diversity of organics in sludge is tremendous. While some are easily degraded, others are persistent. Nonetheless, the EPA concluded in its risk assessment that organics present in sludge present no significant risks as a result of land application. It is instructive to examine this exercise in more detail. The foundation of risk assessment is hazard identification, requiring the development of a complete list of the contaminants of concern [6]. This is somewhat problematic for a material as complex and variable as sludge.

Much of the biosolids contaminant concentration data used in the initial risk assessment effort was obtained from EPA's *Fate of Priority Pollutants in Publicly Owned Treatment Works*, the so-called "40 Cities Study" published in 1982 [7]. Based on this survey and existing data on pollutant toxicities, a list of 200 potential metals and synthetic organics of concern was initially generated by EPA in 1984 [6]. These target compounds represent less than 1 percent of those in commerce. The list was pared to about 50, including 30 individual organic pollutants or groups of compounds (e.g., PCBs) by expert panels. Hazard indices were then determined by comparing expected exposure to a given contaminant via most likely routes (e.g., ingestion of soil) to established human health or other accepted regulatory criteria. Next, hazard rankings were developed based solely on exposure to a contaminant via biosolids alone. Ten organics met or exceeded an index value of 1.0 (exposure/toxicity ratio) in this evaluation. Five additional compounds that had been deleted earlier were later re-added to these ten and subjected to additional scrutiny.

At this juncture of the review process, circa 1989, it was felt that the data from the 1982 "40 Cities Study" was outdated. More stringent industrial pretreatment regulations had taken effect, resulting in generally lower pollutant burdens in influents. Wastewater treatment processes and analytical techniques had also improved. Further, some of the plants examined in the 1982 study had significant industrial input and thus were not representative of the population of WWTPs from which biosolids were typically derived. Finally, the sludges analyzed previously were taken from several stages in the treatment process. Thus they were not representative of the final processed sludge that ultimately serves as the source for the land-applied material. In 1988, the National Sewage Sludge Survey (NSSS) was initiated. It involved 180 facilities and 412 chemicals, chosen in part on the availability of analytical reference standards [6]. This study ultimately concluded that organic pollutant levels in sludge were low. In the end, EPA chose to exclude all organic pollutants from the Part 503 regulations based on the following arguments: 1) persistent pollutants of concern have been restricted or banned from use or manufacture in the United States; 2) the NSSS showed that organic pollutants were not present in sludge at "significant frequencies of detection"; and 3) pollutant concentrations in sludge were less than those determined to be of concern in the risk assessment [6]. Section 405 of the Clean Water Act required EPA to consider a second round of biosolid contaminants for possible regulation. However, a new sludge analysis survey was not conducted and EPA is currently only considering setting limits for chlorinated dioxin-like compounds [8].

Risk assessment is a logical framework for policy development. However, its validity depends on the availability and incorporation of complete data. Recognizing this difficulty and associated time lags, other countries have occasionally used the precautionary principle. Detrimental health effects are a function of all of the chemicals to which organisms are exposed, not just those on regulatory lists or for which detection is convenient using available technology or analytical standards. Most risk assessments also consider only the "worst-case" single chemical exposure, not simultaneous exposure pathways or coincident exposure to multiple chemicals. Recently, it has been observed that simultaneous exposure to several endocrine-disrupting chemicals was capable of generating an observable effect, even though each was present below its respective effects threshold [9]. Some proponents have used the paucity of demonstrated, well-documented detrimental impacts of sludge land application as a rationale not to fill gaps in the database critical to the risk assessment. This is unsatisfying and undermines the contention of the same proponents that risk assessment is superior to alternative strategies as it is based on "sound science." It also transfers the burden of proof from those who benefit from the action to those who might be damaged by an erroneous conclusion as to safety. Regardless, consideration of our past track record of recognizing deleterious effects in the field due to synthetic chemicals is not encouraging. For example, diethylstilbestrol, PCBs, DDT, and chlorofluorocarbons were all assumed to be harmless after initial testing and were used for years before their consequences were fully realized.

One of the conclusions of the EPA risk assessment is that organic pollutants are not present in land-applied sludges at concentrations hazardous to living resources. Obviously, a precondition to this is that the assessor possesses complete data on the identities and quantities of the chemicals actually in sludge. We present information below that suggests this was not the case.

BROMINATED FLAME RETARDANTS (BFRs)

BFRs are current-use synthetic organic compounds. There are several different BFR chemicals in commerce, but here we restrict our attention to the polybrominated diphenyl ethers (PBDEs) due to recent recognition of their importance as chemicals of environmental and human health concern. PBDEs are used in increasing amounts in North America. They are chemically related to PCBs and polybrominated biphenyls (PBBs) and replaced them in some similar applications. PBDE demand in North America, 33,965 metric tons in 1999, represented more than 50 percent of the world's entire consumption [10]. Three formulations, mixtures of PBDE congeners possessing related degrees of bromination, are in commercial production: Deca-BDE (used primarily in thermoplastics and as a back-coating on textiles), Octa-BDE (mostly in thermoplastics and electrical wire coatings), and Penta-BDE (predominantly in polyurethane foam, in turn used in furniture and padding). These products constitute 82 percent, 5.7 percent, and 12.7 percent respectively of the global PBDE market. PBDEs are added to polymers during their formation, but do not covalently bond to them. Thus a small percentage may potentially migrate from the polymer over time. Release to the environment may be greatly accelerated if the polymer itself degrades, as in the case of polyurethane foam [11]. All three PBDE mixtures appear to be environmentally persistent. Penta-BDE is likely the most problematic from a bioaccumulation and toxicological standpoint. Bioconcentration factors for its constituents are comparable to or exceed those of PCBs [12]. About 98 percent of the global demand for Penta-BDE now resides in North America, fostered by well-intended legislation to mitigate the role furniture plays in fires. The Octa- and Deca-BDE mixtures appear to present less of a hazard due to their lower bioavailability and mobility, a result of their extraordinarily low water solubilities and volatilities [13].

Components of the Penta-BDE mixture have now been detected in organisms from even remote areas, for example, marine mammals from the Canadian Arctic [14]. The expected route is via atmospheric transport. PBDEs have been detected in air from both urban and rural areas [15]. In 2001 we reported detectable concentrations (> 5 ug/kg lipid basis) in the edible flesh of nearly 90 percent of the fish samples examined from Virginia waters [16]. Consumption of contaminated fish is believed to be a mechanism for human exposure [17]. PBDEs have also been detected in humans. Until recently, when Penta-PBDE use dropped in Europe, levels in human breast milk in Swedish women were reported to be doubling every five years [18]. The increasing PBDE concentrations were sufficient to trigger European Union action to ban Penta-BDE, effective in 2003. Further assessment of the hazards of Deca-BDE and Octa-BDE mixtures are ongoing. In 2002, it was reported that breast milk from North American women showed a similar increasing trend in PBDE levels [18]. However, concentrations appear to be 40-times higher, consistent with the greater Penta-BDE usage here. Some have suggested that environmental PBDE levels may be driven by decomposition of the more commonly used Deca-BDE product [19]. However, evidence of extensive debromination to Penta-BDE-like congeners in the environment is lacking to date.

Our understanding of the toxicological potential of PBDEs is incomplete and studies to further elucidate their effects are underway. PBDEs appear to have low acute toxicity, but chronic exposure or exposure during development may compromise the endocrine and nervous systems [20, 21]. This is particularly troubling in light of the 2002 report that PBDEs in infant serum were higher than those in other age groups [22]. PBDEs, and even more so their metabolites, are structurally similar to the thyroid hormone thyroxine (T4) (Figure 1). Some have been observed to displace T4 from the plasma transport protein transthyretin *in-vitro* [23]. Little information on effects of PBDEs on humans exists. However, the structurally related PBBs have been associated with precocious pubertal events in exposed girls [24].

ALKYLPHENOL POLYETHOXYLATES (APEOs)

APEOs are predominantly used as surfactants in heavy-duty detergents. The nonylphenol polyethoxylates (NPEOs) constitute about 80 percent of the APEO market, with octylphenol analogs (OPEO) constituting the bulk of the remainder [25]. Interestingly, NPEOs have also been used as human spermicides. In 2000, U.S. demand for NPs, mainly as NPEO precursors, was 109 million kg [26].

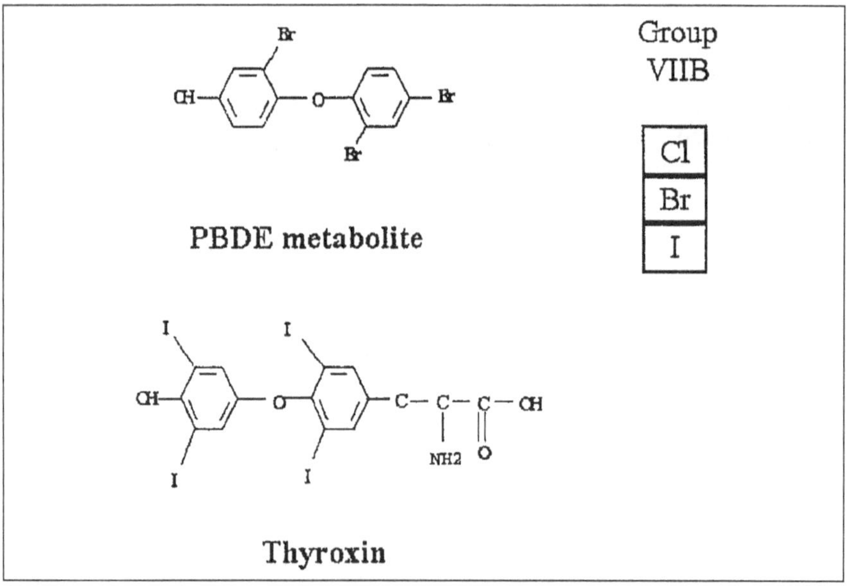

Figure 1. PBDE metabolites bear a striking resemblance to the thyroid hormone thyroxin and can compete with it for binding sites on the transport protein transthyretin. The halogens (Cl, Br and I) all belong in Group VIIB of the periodic chart. However, Br is more similar to I, the latter contained in thyroxin, than Cl. PBDEs are identical to PBBs except for the presence of the ether linkage in the former.

Growth of U.S. demand for NP has been about 2 percent per year since 1996 and NPEO exports doubled to 22 million kg from 1999 to 2000. In contrast, several European countries, e.g., Norway, Switzerland, and Denmark, have reduced their use of APEOs in light of the detection of part-per-million NP concentrations in aquatic sediments and sludges and elucidation of their toxic and endocrine-disruptive properties. NPs are acutely toxic to some aquatic organisms (< 10 μg/l water column or 315 mg/kg in sediments) [27, 28] and have been linked to impacts on some endocrine-related processes at less than 10 μg/L [29]. Less is known regarding effects in terrestrial systems.

APEOs themselves have relatively low toxicities and generally are believed to degrade in the environment in the presence of oxygen. However, a fraction of the APEOs may be incompletely degraded to the corresponding alkylphenols (APs: mostly NPs and OP), especially under low oxygen conditions, such as those present in some WWTPs [30]. APs are moderately hydrophobic and partition preferentially to particulates and thus accumulate in sediments and sewage sludge. In testament to this pattern, we observed NP burdens of 54,400 μg/kg in York River surficial sediments, near a WWTP that ceased operation more than 20 years previously [31]. A half-life in excess of 60 years was calculated for APEOs in cold, anoxic sediments from British Columbia [32]. Obviously, consideration of the environment to which pollutants are released must be factored in when estimating their persistence.

Uptake and retention of NPs and OP by organisms may explain their reported greater estrogenicity *in vivo* compared to *in vitro* assays [33]. In the field, reproductive disturbances in wild fish populations have been observed near WWTP outfalls, including intersexuality, vitellogenesis, and decreased gonado-somatic index (ratio of gonadal size to body weight) [34]. In all likelihood, these impacts are a result of exposure to a combination of endocrine-disrupting chemicals. In addition, applications of NP-containing insecticides on Canadian forests have been implicated in declines of Atlantic salmon returns, possibly due to impacts on developing smolt [35]. These data demonstrate that exposure in the field to estrogenic chemicals has deleterious consequences to wildlife.

SURVEY OF PBDEs AND APs IN U.S. BIOSOLIDS

We examined sewage sludges generated by WWTPs located in six different U.S. states for APs, PBDEs, and related compounds (Table 1). All biosolids had been subjected to additional stabilization procedures in preparation for land application, including: liming, composting, anaerobic digestion, and heat treatment. Samples selected included Class A (low pathogens) and Class B (application restrictions due to pathogen burdens) biosolids. Analytical procedures have been described elsewhere [36, 37]. Briefly, samples were lyophilized, surrogate standards added, extracted by enhanced solvent extraction and purified by adsorptive and size exclusion liquid chromatography. Compounds were identified

Table 1. Total AP, Penta-BDE, and Deca-BDE Concentrations (μg/kg Dry Weight Basis) in the U.S. Biosolids Examined

Location (U.S. state)[a]	Biosolid stabilization technique	Biosolid class	% total organic carbon	Total AP + NP1EOs + NP2EOs[b,c]	Total Penta-like PBDEs	Deca-PBDE
VA	compost	A	9.90	6,100	1500	308
VA	compost	A	18.5	176,000	2290	1460
TX	compost	A	16.1	14,200	1290	368
VA	lime	B	12.3	932,000	1110	553
VA	lime	B	24.6	529,000	1660	84.8
NY/MD	heat	A	24.9	544,000	1500	1940
NJ	heat	A	32.0	NA	2110	4890
CA	anaerobic	B	23.5	721,000	1430	347
CA	anaerobic	B	22.2	758,000	1100	340
CA	anaerobic	B	25.4	925,000	1590	450
CA	anaerobic	B	20.6	768,000	1620	389
CA	anaerobic	B	28.8	981,000	2010	NA
Mean			21.6	578,000	1600	1010
Median			22.9	721,000	1550	389
Std. Dev.			6.48	362,000	375	1400
CV			30.1%	62.6%	23.4%	139%

[a]CV = coefficient of variability. [b]NA = analyte not determined in this sample. [c]Total includes NPs, OP, and the NP mono- and diethoxylates.

and quantified by gas chromatography (GC) with mass spectrometric (MS) and electrolytic conductivity detection in the halogen selective mode (ELCD-HSM). A major strength of modern analytical procedures is their selectivity and sensitivity. For this reason, MS in the selected ion mode (SIM) is increasingly used in environmental studies. In SIM-MS only specific ion fragments of the desired target compound(s) are monitored and recorded. Compounds not exhibiting these fragments upon ionization generate no response. This avenue provides a very specific and accurate measurement capability, even when considerable numbers/amounts of co-extracted compounds are present. The downside is that other pollutants present in the sample may be overlooked. This may explain the failure to detect the NPs and PBDEs in previous EPA sludge surveys. Figure 2 shows a chromatogram from the GC/ELCD-HSM analysis of a sludge sample. The Penta-BDE related congeners (BDE-47, 99, 100, 153, and 154) are obvious here, dwarfing the PCBs (most abundant congeners: PCB-153 and 138), even

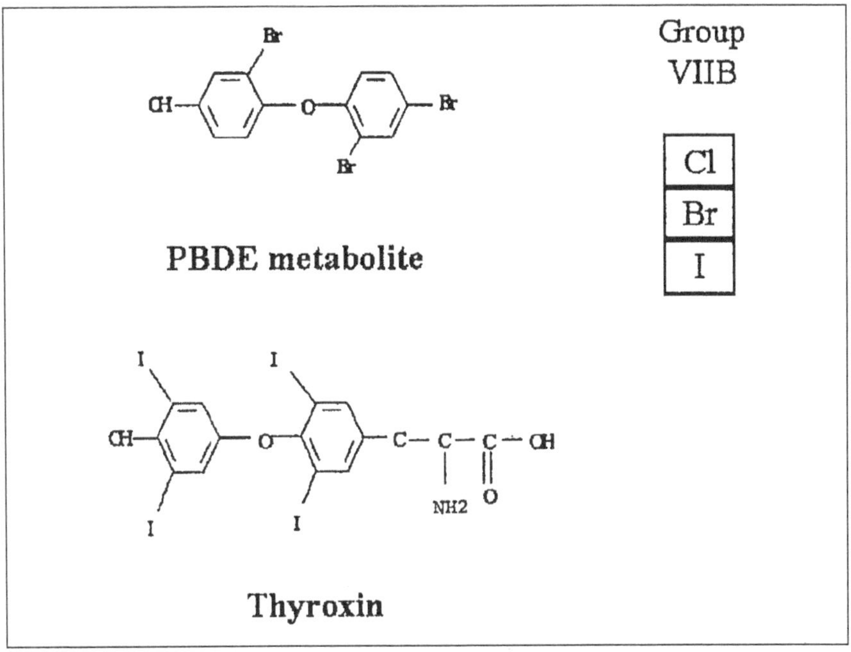

Figure 2. Chromatogram of a biosolid extract using the halogen selective ELCD. The largest peaks observed belong to the PBDEs (i.e. BDE-47 and 99), rather than to the chlorinated pollutants (e.g., PCBs and chlordanes) considered by EPA in their risk assessment. Note that the NSSS reported a detection frequency of 0 percent for chlordane. istd = internal standard added to the sample for quality control and quantification purposes.

though this detector is about two-fold more sensitive to chlorine than to bromine. Interestingly, the NSSS failed to detect chlordanes in any sludges. Based on this and the fact that chlordane use was banned in the United States in 1988, EPA removed them from further consideration for regulation under Part 503. However, we encountered chlordanes at easily detectable levels in modern biosolids (see Figure 2). Note that Deca-BDE (BDE-209) is not shown in Figure 2 due to its long GC retention time. A separate GC run with a shorter column is required to elute Deca-BDE due to its low volatility.

Several aspects of our PBDE sludge data merit emphasis (Table 1). First, the relative abundances of the individual congeners present closely resembled those in the commercial Penta-BDE product. This suggests that PBDE input to sludge may be derived somewhat directly from the original products they were used in, as major shifts in relative congener abundances due to differences in physical properties, such as water solubility, had not occurred. Additionally, Penta-BDE concentrations were 10-40 times greater than those in European sludges, proportional to the greater North American demand for this product [36]. Penta-BDE levels were also remarkably consistent between sludges. This suggests these PBDEs may be derived from discarded products rather than from PBDE or polyurethane foam production facilities, as manufacturers are heterogeneously distributed in the U.S. If true, improved industrial pretreatment regulations would not affect these inputs. Penta-BDE concentrations were similar as a function of the sludge stabilization processes, not surprising considering their persistence. In contrast, Deca-BDE concentrations varied substantially between sludges, ranging from 84.8 to 4890 μg/kg (coefficient of variability (CV): 139 percent). In contrast, total Penta-BDE concentrations ranged from 1100 to 2290 μg/kg (CV: 23.4 percent). This may relate to Deca-BDE's predominant use in environmentally stable thermoplastics and its extremely low water solubility and volatility. In contrast, the surface of polyurethane foam can degrade into a powder when exposed to outdoor weathering. This dust, which may contain 10-30 percent by weight Penta-BDE, is easily transportable by wind and water runoff. We estimate that a single medium-sized (1 kg) seat cushion, that has been flame retarded with Penta-BDE (conservatively at 10 percent by weight), contains enough PBDE to contaminate 100,000 kg (dry weight) of biosolids to a concentration of 1000 μg/kg by weight, comparable to the mean Penta-BDE level (1560 μg/kg) we observed [11].

The U.S. sludges examined also contained up to 981 mg/kg of APs (total of OP, NPs, plus the NP mono- and diethoxylates, intermediate degradates of NPEOs) (Table 1). Total NP concentrations varied as a function of the sludge stabilization process. As NPs are generated from NPEOs under anaerobic conditions, it is not surprising that the anaerobically stabilized sludges had some of the highest total NP concentrations. However, relatively high levels were present in the limed and heat-treated sludges as well. NPs are vulnerable to aerobic degradation and the composted sludges had the lowest concentrations. OP was detected in nine of the 11 biosolids and also peaked in the anerobically digested samples, but levels were typically less than 2 percent of NP concentrations.

Total NPs concentrations in all but one U.S. sludge exceeded the 10 mg/kg Danish limit for land-applied biosolids. In recent years NP burdens in Danish sludge have decreased to an average of 4 mg/kg due to reduced NPEO usage [38]. A soil Estimated No Effects Value of 0.34 mg/kg has been recommended by Environment Canada, based on potential impacts on earthworm reproduction. Assuming a biosolid application factor of 0.003 (3 tons dry biosolid to 1 acre soil, tilled to 15 cm depth), all but two of the U.S. biosolids examined would exceed this recommendation when applied [37]. Biosolid application rates on nonagricultural lands (e.g., parks, forests and reclamation sites) may substantially exceed this loading. For example, application of 35 tons per acre was permitted at the Stafford Regional Airport in northern Virginia in 2002. Sludge-associated APs have been reported to be degraded in a matter of weeks in aerobic soils. However, in soil aggregates entry of oxygen may be impeded and NPs may persist [39]. NPs have been detected in soil cores below the surface for considerable periods after application [40]. The higher ethoxylates and carboxylic acid derivatives would exhibit more mobility due to their greater water solubility. NPs show moderate partitioning to soil, particularly to the organic fraction. Preferential flow via passages in the soil (e.g., insect or animal burrows, shrink/swell fractures . . .) can also enhance migration, as may the presence of natural or synthetic organic matter [5]. The NPs themselves are derived from surfactant parent compounds. Other surfactants, such as linear alkylbenzene sulfonates are also common sludge constituents [5]. NPs have also been detected in groundwater as a result of infiltration from septage drainage fields [41]. In a study of contaminants in secondary treated sewage effluent introduced into a shallow unconfined aquifer near Boston, Massachusetts (USA), NP mobility in the groundwater was significant, but less than more water-soluble organics [42]. However, degradation of the NPs in the groundwater was deemed minimal.

In addition to the APs and PBDEs, we observed the presence of numerous other organic pollutants. Some of them were reported in the NSSS, including polycyclic aromatic hydrocarbons, PCBs, DDT degradation products, and chlordanes. They were considered singly in the EPA risk assessment process, but additive effects may also be possible. In addition, we observed other organic contaminants in the sludge that were not covered in the EPA risk assessment, including synthetic musk compounds, triclosan and tributytin (unpublished data). Musks are commonly used as fragrances and some have been found to be bioaccumulative [1]. Triclosan is an anti-bacterial agent, used increasingly in household cleaning products and toothpaste. Tributyltin is a biocide and anti-foulant. It has seriously impacted some coastal shellfish populations and is highly toxic to crustaceans. Additional contaminants already detected in WWTP effluents are most certainly present in sludge, including pharmaceuticals, such as ethinylestradiol, the active ingredient in birth control pills.

Improved product stewardship is obviously indicated for biosolids. A recent EPA Inspector General report was highly critical of EPA's oversight of the biosolids program [43]. The concern of the biosolids industry and EPA over organic contaminants also contrasts with that of the composting industry. The latter reacted strongly to reports of the presence of a single herbicide, clopyralid, in U.S. compost [44]. Clopyralid's presence was due to its initial application on lawns and related uses and its subsequent transfer with clippings to compost. While the herbicide has not been reported in sewage sludge, compost derived from waste vegetation may be added to sludge during stabilization. Since its discovery, and observation of associated detrimental effects on plants, individuals in the compost industry have proposed that the manufacturer of clopyralid, Dow AgroSciences, be held accountable for all damages associated with the use of contaminated compost [45]. In the case of biosolids, the producer/dischargers of the contaminants are essentially anonymous. Precautions such as more extensive re-labeling of clopyralid by Dow to warn applicators that its use precludes subsequent composting of clippings was deemed insufficient to curtail its entry into the compost wastestream. In the case of biosolids, chemicals are intentionally released to the wastestream with no concern regarding their impact on the quality of biosolids and knowledge of the identities of the contaminants therein is incomplete. The composting industry has requested an independent investigation of the extent of clopyralid contamination at composting facilities. EPA has deemed an updated survey of contaminants in biosolids to be unnecessary. Suspension of registration of new clopyralid-containing products and of existing applications in affected geographical areas has been suggested. Consideration of the possible occurrence of herbicide residues in compost as a future requirement for governmental registration of all pesticides has surfaced. In contrast, regulation of the release of many biosolid contaminants, e.g., PBDEs, APs, and pharmaceuticals, to WWTPs is unregulated. Calls have also been made to require Dow to compensate compost facilities and downstream users for real or potential clopyralid-contaminated compost damages, including the cost of land remediation. In the case of biosolids, contamination of land in the United States is expected and is deemed acceptable to "maximum" limits.

Inadequate information on the identities and concentrations of synthetic organic pollutants in sludge has been identified as a major data gap by others, for example, in a 2002 European Union review [46]. Previously, a 1996 National Research Council report questioned EPA's exemption of organic pollutants from the Part 503 sludge rule and indicated that more complete data on the range and concentrations of organic pollutants was essential [47]. The presence of environmentally persistent chemicals, e.g., PBDEs, in land-applied sludge is certainly problematic. However, persistence is a function of ambient conditions. Also, detrimental effects on- or off-site may occur within the lifetime of even short-lived chemicals. Thus, perhaps persistence should not be an essential criterion for a chemical of concern, except in the context of tightly controlled waste land-farming scenarios. Our understanding of the potential detrimental effects of sludge-associated chemicals is limited. Impacts due to multiple interacting chemicals and to an expanded array of potential modes of toxicity, such as endocrine disruption, merit additional investigation. For example, recently sludge extracts have been demonstrated to be estrogenic in *in-vitro* tests [48]. It is now known that current-use organic contaminants of toxicological concern are indeed present at high concentrations in land-applied sludge. Some may be entering through the use of consumer products, not industrial sources. Therefore, more stringent industrial pretreatment regulations may have no effect on releases of them. In conclusion, the above results indicate that the premises utilized by EPA to exempt synthetic organics in land-applied sewage sludge from regulation under Part 503 are questionable.

AUTHOR'S NOTE

Since submission of this manuscript, the National Research Council, at the request of the US EPA, completed its second review of the technical basis of the biosolids chemical and pathogen regulations [49]. The review concluded that, to date, there was no documented scientific evidence that the Part 503 rule has failed to protect public health. However, it also stated that the applicable epidemiological research conducted has been inadequate. The report also noted the inadequacy of the 1988 NSSS, the need for an updated survey of chemicals in biosolids, as well as to more fully consider exposures and risks associated with the extensive suite of organic chemicals (including PBDEs and NPs) in biosolids.

This article is Contribution No. 2533 of the Virginia Institute of Marine Sciences, The College of William and Mary.

REFERENCES

1. R. C. Hale and M. J. La Guardia. Emerging Contaminants of Concern in Coastal and Estuarine Environments. Chapter 3 in *Risk Assessment in Coastal and Estuarine Environments,* M. C. Newman, M. H. Roberts Jr., and R. C. Hale (eds.), pp. 41-72, 2002.
2. U.S. EPA, *Biosolids Generation, Use, and Disposal in the United States, EPA530-R-99-009,* 1999.
3. *Municipal Solid Waste in the United States 1999 Facts and Figures.* http://www.epa.gov/epaoswer/non-hw/muncpl/msw99.htm.
4. U.S. EPA, *The Standards for the Use or Disposal of Sewage Sludge, Title 40 of the Code of Federal Regulations, Part 503,* 1993.
5. E. Z. Harrison, M. B. McBride, and D. R. Bouldin, *Working Paper: The Case for Caution: Recommendations for Land Application of Sewage Sludges and an Appraisal of the U.S. EPA's Part 503 Sludge Rules,* Cornell Waste Management Institute, 1999.
6. U.S. EPA, *A Guide to the Biosolids Risk Assessments for the EPA Part 503 Rule,* Office of Wastewater Management, U.S. EPA/ 8332//B-93-005, 1995.
7. U.S. EPA, *Fate of Priority Pollutants in Publicly Owned Treatment Works, EPA-440/1-82-303,* 1982.
8. U.S. EPA, *Proposed Rule Revising the Standards for Use and Disposal of Biosolids, EPA-822-F-99-005,* 1999.
9. E. Silva, N. Rajapakse, and A. Kortenkamp, Something from "Nothing"—Eight Weak Estrogenic Chemicals Combined at Concentrations Below NOECs Produce Significant Mixture Effects, *Environmental Science & Technology, 36,* pp. 1751-1756, 2002.
10. Major Brominated Flame Retardants Volume Estimates, *Brominated Science and Environmental Forum,* http://www.bsef.com, 2001.
11. R. C. Hale, M. J. La Guardia, E. Harvey, and T. M. Mainor, The Potential Role of Fire Retardant-Treated Polyurethane Foam as a Source of Brominated Diphenyl Ethers to the U.S. Environment, *Chemosphere, 46,* pp. 729-735, 2002.

12. J. Peltola and L. Yla-Mononen, *Pentabromodiphenyl Ether as a Global POP.* TemaNord Nordic Council of Ministers, 2001.
13. M. L. Hardy, The Toxicology of the Three Commercial Polybrominated Diphenyl Oxide (Ether) Flame Retardants, *Chemosphere, 46,* pp. 757-777, 2002.
14. M. G. Ikonomou, S. Rayne, and R. F. Addison, Exponential Increases of the Brominated Flame Retardants, Polybrominated Diphenyl Ethers, in the Canadian Arctic from 1981-2000, *Environmental Science & Technology, 36,* pp. 1886-1892, 2002.
15. B. Strandberg, N. G. Dodder, I. Basu, and R. A. Hites, Concentrations and Spatial Variations of Polybrominated Diphenyl Ethers And Other Organohalogen Compounds in Great Lakes Air, *Environmental Science & Technology, 35,* pp. 1078-1083, 2001.
16. R. C. Hale, M. J. La Guardia, E. P. Harvey, T. M. Mainor, W. H. Duff, and M. O. Gaylor, Polybrominated Diphenyl Ether Flame Retardants in Virginia Freshwater Fishes (USA), *Environmental Science & Technology, 35,* pp. 4585-4591, 2001.
17. A. Sjodin, L. Hagmar, E. Klasson-Wehler, J. Bjork, and A. Bergman, Influence of the Consumption of Fatty Baltic Sea Fish on Plasma Levels of Halogenated Environmental Contaminants in Latvian and Swedish Men, *Environmental Health Perspectives, 108,* pp. 1035-1041, 2000.
18. K. S. Betts, Rapidly Rising PBDE Levels in North America, *Environmental Science & Technology, 36,* pp. 50A-52A, 2002.
19. R. Renner, What Fate for Brominated Fire Retardants? *Environmental Science & Technology, 34,* pp. 223-226A, 2000.
20. P. O. Darnerud, G. S. Eriksen, T. Johannesson, P. B. Larsen, and M. Viluksela, Polybrominated Diphenyl Ethers: Occurrence, Dietary Exposure, and Toxicology, *Environmental Health Perspectives, 109,* pp. 49-68, 2001.
21. P. Eriksson, E. Jakobsson, and A. Fredriksson, Brominated Flame Retardants: A Novel Class of Developmental Neurotoxicants in Our Environment, *Environmental Health Perspectives, 109,* pp. 903-908, 2001.
22. C. Thomsen, E. Lundanes, and G. Becher, Brominated Flame Retardants in Archived Serum Samples from Norway: A Study on Temporal Trends and the Role of Age, *Environmental Science & Technology, 36,* pp. 1414-1418, 2002.
23. I. Meerts, J. J. van Zanden, E. A. C. Luijks, I. van Leeuwen-Bol, G. Marsh, E. Jakobsson, A. Bergman, and A. Brouwer, Potent Competitive Interactions of Some Brominated Flame Retardants and Related Compounds with Human Transthyretin in Vitro, *Toxicological Science, 56,* pp. 95-104, 2000.
24. H. M. Blanck, M. Marcus, P. E. Tolbert, C. Rubin, A. K. Henderson, V. S. Hertzberg, R. H. Zhang, and L. Cameron, Age at Menarche and Tanner Stage in Girls Exposed in Utero and Postnatally to Polybrominated Biphenyl, *Epidemiology, 11,* pp. 641-647, 2000.
25. T. Isobe, H. Nishiyama, A. Nakashima, and H. Takada, Distribution and Behavior of Nonylphenol, Octyphenol, and Nonylphenol Monoethoxylate in Tokyo Metropolitan Area: Their Association with Aquatic Particles and Sedimentary Distributions, *Environmental Science & Technology, 35,* pp. 1041-1049, 2001.
26. Chemical Profile: Nonylphenol, *Chemical Market Reporter, 260,* http://www.aperc.org/nonylphenol.pdf, 2001.
27. C. G. Naylor, Environmental Fate and Safety of Nonylphenol Ethoxylates, *Textile Chemist & Colorist, 27,* pp. 29-33, 1995.

28. R. Bettinetti, D. Cuccato, S. Galassi, and A. Provini, Toxicity of 4-Nonylphenol in Spiked Sediment to Three Populations of Chironomous Riparius, *Chemosphere, 46,* pp. 201-207, 2002.
29. S. R. Miles-Richardson, S. L. Pierens, K. M. Nichols, V. J. Kramer, E. M. Snyder, S. A. Snyder, J. A. Render, S. D. Fitzgerald, and J. P. Giesy, Effects of Waterborne Exposure to 4-Nonylphenol and Nonylphenol Ethoxylate on Secondary Sex Characteristics and Gonads of Fathead Minnows (Pimephales Promelas), *Environmental Research, 80,* pp. 122-137, 1999.
30. M. Ahel, W. Giger, and M. Koch, Behavior of Alkylphenol Polyethoxylate Surfactants in the Aquatic Environment—1. Occurrence and Transformation in Sewage Treatment, *Water Research, 28,* pp. 1131-1142, 1994.
31. R. C. Hale, C. L. Smith, P. O. de Fur, E. Harvey, E. O. Bush, M. J. La Guardia, and G. G. Vadas, Nonylphenols in Sediments and Effluents Associated with Diverse Virginia (USA) Wastewater Outfalls, *Environmental Toxicology and Chemistry, 19,* pp. 946-952, 2000.
32. D. Y. Shang, R. W. MacDonald, and M. G. Ikonomou, Persistence of Nonylphenol Ethoxylate Surfactants and Their Primary Degradation Products in Sediments Obtained from Near a Municipal Outfall in the Strait of Georgia, British Columbia, Canada, *Environmental Science & Technology, 33,* pp. 1366-1372, 1999.
33. A. M. Ferreira-Leach and E. M. Hill, Bioconcentration and Distribution of 4-Tert-Octylphenol Residues in Tissues of the Rainbow Trout (Oncorhynchus mykiss), *Environmental Research, 51,* pp. 75-89, 2001.
34. S. Jobling, M. Nolan, C. R. Tyler, G. Brighty, and J. P. Sumpter, Widespread Sexual Disruption in Wildfish, *Environmental Science & Technology, 32,* pp. 2498-2506, 1998.
35. W. L. Fairchild, E. O. Swansburg, J. T. Arsenault, and S. C. Brown, Does an Association between Pesticide Use and Subsequent Declines in Catch of Atlantic Salmon (Salmo salar) Represent a Case of Endocrine Disruption? *Environmental Health Perspectives, 107,* pp. 349-358, 1999.
36. R. C. Hale, M. J. La Guardia, E. Harvey, M. O. Gaylor, T. M. Mainor, and W. H. Duff, Flame Retardants: Persistent Pollutants in Land-Applied Sludges, *Nature, 412,* pp. 141-142, 2001.
37. M. J. La Guardia, R. C. Hale, E. Harvey, and T. M. Mainor, Alkylphenol Ethoxylate Degradation Products in Land Applied Sewage Sludges (Biosolids), *Environmental Science & Technology, 35,* pp. 4798-4804, 2001.
38. *Danish Environment & Energy Newsletter*, No. 11, Ministry of the Environment, 2001.
39. M. Hesselse, D. Jensen, K. Skals, T. Olesen, P. Moldrup, P. Roslev, G. Krog Mortensen, and K. Henriksen, Degradation of 4-Nonylphenol in Homogeneous and Nonhomogeneous Mixtures of Soil and Sewage Sludge, *Environmental Science & Technology, 35,* pp. 3695-3700, 2001.
40. J. Vikelsoe, M. Thomsen, E. Johansen, and L. Carsen, *Phthalates and Nonylphenols in Soil: A Field Study of Different Soil Profiles,* NERI Technical Report No. 268, Ministry of Environment and Energy, National Environmental Research Institute, Denmark, 1999.
41. R. A. Rudel, S. J. Melly, P. W. Geno, G. Sun, and J. G. Brody, Identification of Alkylphenols and Other Estrogenic Phenolic Compounds in Wastewater, Septage, and Groundwater on Cape Cod, Massachusetts, *Environmental Science & Technology, 32,* pp. 861-869, 1998.

42. L. B. Barber II, E. M. Thurman, and M. P. Schroeder, Long-Term Fate of Organic Micropollutants in Sewage-Contaminated Groundwater, *Environmental Science & Technology*, *22*, pp. 205-211, 1988.
43. U.S. EPA, *Office of Inspector General, Status Report: Land Application of Biosolids, Report No. 2002-S-000004,* 2002
44. R. Rynk, Industries Respond to the Clopyralid Controversy, *BioCycle, 42,* pp. 66-67, 2001.
45. *U.S. Composting Council Clopyralid Position Paper,* October 30, 2001.
46. *Disposal and Recycling Routes for Sewage Sludge,* Synthesis Report for the European Commission DG Environment-B/2, 2002.
47. *Use of Reclaimed Water and Sludge in Food Crop Production,* National Academy Press, Washington DC, 1996.
48. K. Rehmann, K.-W. Schramm, and A. A. Kettrup, Applicability of a Yeast Oestrogen Screen for the Detection of Oestrogen-Like Activities in Environmental Samples, *Chemosphere, 38,* pp. 3303-3312, 1999.
49. *Biosolids Applied to Land: Advancing Standards and Practice,* National Academy Press, Washington, DC, 2002.

Note: Originally published in *New Solutions, 12*(4), pp. 371-386, 2002.

CHAPTER 11

Are We Winning or Losing the War on Cancer? Deciphering the Propaganda of NCI's 33-Year War

Genevieve K. Howe and Richard W. Clapp

The National Cancer Institute (NCI) and collaborating agencies have proclaimed great progress in the U.S. "war on cancer," while at the same time presenting more reasons for concern than celebration. We reviewed various documents and data files and found that incidence and mortality rates for all cancer sites combined remain higher than they were when the "war on cancer" was declared in 1971, despite very recent, modest decreases. The burden of the disease has risen from three million to nearly ten million people. Black Americans, men of all races, and other segments of the population disproportionately bear the burden of cancer. We also looked at data for malignant breast cancer and found that incidence rates increased 36% from 1973 to 2000, while mortality for all population groups combined declined slightly. Breast cancer mortality is 34% higher among black women than among white women, even though white women are generally more likely to get the disease. The $50 billion spent on the "war on cancer" over the last 33 years has yielded few gains. The NCI's resources must be refocused on preventing cancers we know how to prevent.

Is the war on cancer over yet? Did we win? Can we send the troops home, bury the last victims, declare the "survivors" cured, and celebrate victory? That depends on what you read and whom you believe. In what follows, we update observations on cancer trends we made in this journal in 1997.[1]

[1] R. W. Clapp, The Decline in U.S. Cancer Mortality from 1991 to 1995. *New Solutions, 7*(4), 1997.

There is no dispute over government data predicting that more than 1.3 million Americans will be diagnosed with some form of cancer[2] and nearly 556,500 will die of cancer *this year*. And, it's been written on the wall for the last 40 years that cancer is the second leading cause of death after heart disease. One in four of all deaths in the United States is now due to cancer [1] and cancer has become the leading cause of death for people between the ages of 35 and 74 [2].

Despite claims by the National Cancer Institute (NCI) that "there have been incredible advances in cancer detection, prevention, and treatment" [3] and that "we are making extraordinary progress in cancer research" [4], the age-adjusted death rate per 100,000 people due to all cancer sites combined rose steadily during the 20th century until its final decade. From 1990 to 2000, the overall cancer mortality rate decreased by 7.6%– prompting many of the recent claims of success. Nonetheless, the death rate for all cancer sites in the year 2000 was 74% higher than it was in 1900 and 0.5% higher than it was in 1970 shortly before President Richard Nixon launched his "war on cancer." The recent small decline in overall cancer mortality, which is largely due to recent declines in male lung cancer mortality, has brought the overall cancer mortality rate back to where it was in the late 1940s [5].

Nearly one out of every two men and more than one out of every three women will receive a cancer diagnosis during their lifetimes [6].

WE'RE WINNING THE WARS ON HEART DISEASE AND STROKES—BUT NOT ON CANCER

Yet, we are winning the wars with most other leading causes of death. Mortality rates due to influenza and pneumonia, the leading cause of death in 1900, decreased by 92% during the 20th century. The rate of death due to cerebrovascular diseases (namely strokes) decreased by 75% and the death rate due to accidents (unintentional injuries) decreased by 61%. Tuberculosis dropped from being the fourth leading cause of death in 1900 to a negligible cause of death. Despite steeply rising and then declining rates over the course of the century, even mortality rates for diseases of the heart, the leading cause of death ever since 1910, decreased slightly (3%) from 1900 to 2000 (Figure 1) [7].

The 20th century saw an increase in the death rates of diabetes mellitus and the additions of chronic lower respiratory diseases, Alzheimer's disease, and septicemia to the ten leading causes of death in the United States [8]. Yet, these causes together with kidney disease accounted for only 13% of all age-adjusted deaths in the year 2000 [9].

[2] All cancer data cited in this chapter exclude non-melanoma skin cancers.

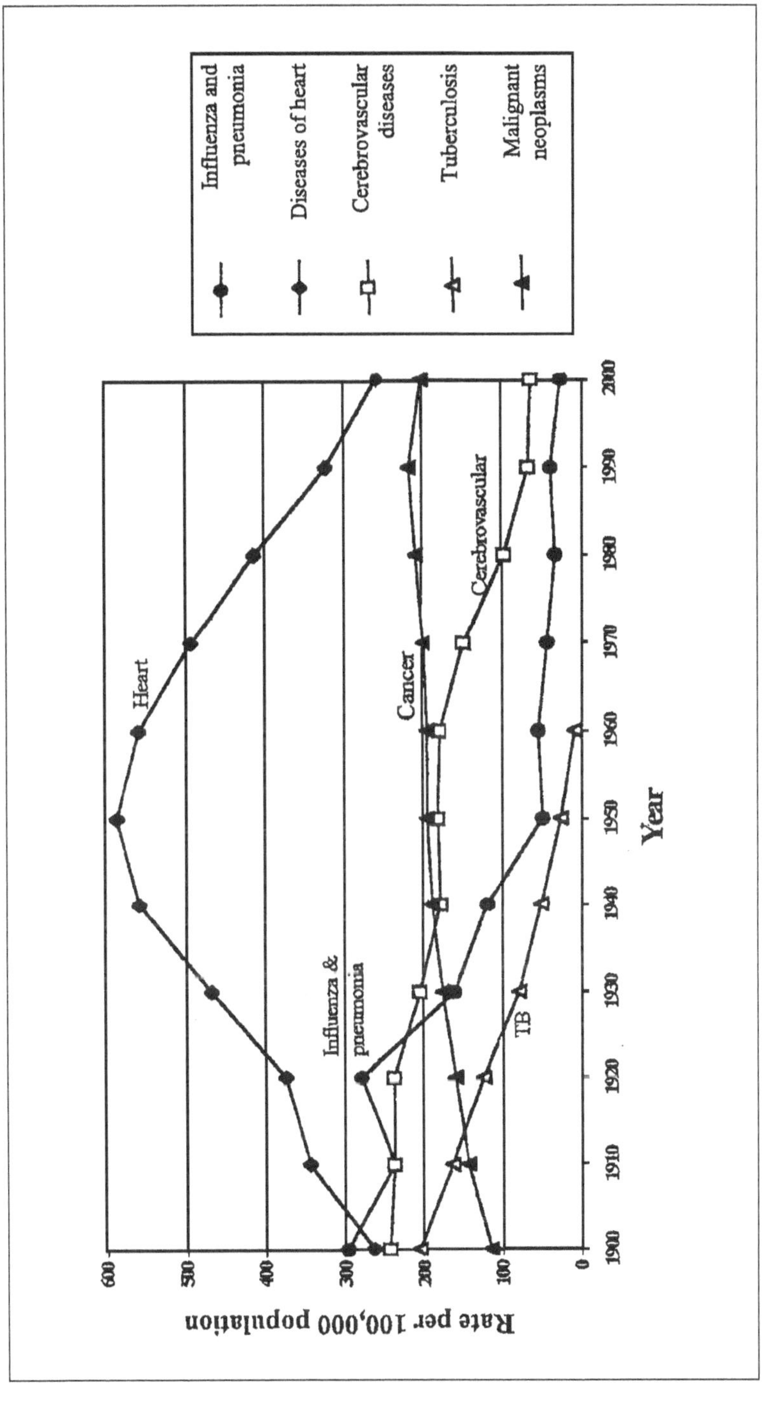

Figure 1. Mortality rates for leading causes of death in the United States for selected years, 1900-2000, per 100,000 population [41].

THE "WAR ON CANCER," $50 BILLION AND 33 YEARS LATER

Cancer became the second leading cause of death in the United States, after diseases of the heart, in the early 1960s. The mounting concern over increasing cancer incidence and mortality rates prompted President Nixon to declare his "war on cancer" in 1971. Since then, we have devoted more than $50 billion in federal taxes to this war, counting only the ever-larger annual appropriations to the NCI. NCI's annual budget grew more than twenty-fold from $233 million in fiscal year 1971 to $4.8 billion in fiscal year 2004 [10]. NCI has requested $6.2 billion for fiscal year 2005 [11].

In December 1971, Nixon signed into law the National Cancer Act, in which Congress declared: "The incidence of cancer is increasing and cancer is the disease which is the major health concern of Americans today." The act significantly enhanced the research capacities of the NCI, while doing little to prevent people from getting cancers with known or suspected causes. The act's primary function was to establish ". . . fifteen new centers for clinical research, training, and demonstration of advanced diagnostic and treatment methods relating to cancer." The act described in only one sentence how new "Cancer Control Programs" would receive 5% (or $20 million) of the first year's funding to establish ". . . programs as necessary for cooperation with State and other health agencies in the diagnosis, prevention, and treatment of cancer" [12]. Prevention programs were usually limited to individual lifestyle changes such as: quit smoking, eat a lower fat diet, and exercise more.

Dr. Samuel Epstein analyzed the NCI's budget in 1993 and found that the agency devoted only 1% of its resources to research on primary cancer prevention and occupational cancer [13].

What progress have we made since 1971? In 1973, the first year of operation for NCI's ambitious Surveillance Epidemiology and End Results (SEER) database, 385 of every 100,000 Americans were newly diagnosed with cancer, and 199 of every 100,000 Americans died from cancer [14]. In the year 2000, 473 of every 100,000 Americans were newly diagnosed with cancer (a 23% increase) and 199.6 of every 100,000 Americans died of cancer (ever so slightly above the 1973 rate) [15].

The $50 billion allocated to the NCI since 1971 has led to advances in cancer therapies, which to varying degrees have somewhat improved the odds of surviving most types of cancer. Certainly, each untimely cancer death avoided is cause for celebration. But, the rates at which people are now getting cancer have increased so significantly that the suffering, psychological, and economic burdens of living with cancer are far greater than ever before. And, because our population is larger and older, the absolute numbers of people getting sick and dying from cancer have increased enormously—leveling off only in the last few years.

According to the NCI, "Cancer-related costs account for about 10 percent of the total amount spent on disease treatment in the United States. Cancer is a major national burden." NCI and its parent agency, the National Institutes of Health, estimated that the overall cost of cancer to the United States in the year 2000 was $180.2 billion: $60 billion for direct medical costs, $15 billion for indirect morbidity costs (the cost of lost productivity due to illness), and $105.2 billion for indirect mortality costs (the cost of lost productivity due to premature death) [16]. If we add the $3.3 billion budget of the NCI in 2000, the cost would become $183.5 billion. In either case, it's a figure far greater than the $108 billion estimated for the first full year of President George W. Bush's war on Iraq [17].

There are other signs that things are not getting better, particularly for certain segments of the population. The National Center for Health Statistics' *Health, United States, 2003* report says that although incidence rates for all cancer sites combined declined in the 1990s for males (the largest contributor being the decline in smoking-related lung cancers), "there was no significant change in cancer incidence for females overall," and that "breast cancer is the most frequently diagnosed cancer among females," especially for non-Hispanic white females. For black Americans, overall cancer death rates were 25% higher than for whites in 2001, with black men suffering the highest toll at 331 per 100,000 [18].

ELIMINATE THE SUFFERING AND DEATH DUE TO CANCER BY 2015?

In 2001, the NCI adopted an astonishing, bold objective: "*Eliminate the suffering and death due to cancer by 2015.*" NCI Director Andrew C. von Eschenbach closed his cover letter to the FY 2005 budget proposal with, "We will not fail!" He also wrote, "Over the past decade, Americans have experienced a 7 percent decline in mortality from cancer and hundreds of thousands of lives have been saved" [19]. Curiously, the same document, at the beginning of a section entitled "Highlights of Progress," states:

> One in four deaths in the United States is attributable to cancer, and one in three Americans will eventually develop some form of cancer. Each day, 3,400 people in America are diagnosed with cancer and another 1,500 die from the disease. But the burden of cancer is too often greater for the poor, ethnic minorities, and the uninsured than for the general population [20].

The NCI doesn't sound optimistic here and it's hard to tell which part of the above paragraph is a sign of "progress." The NCI sounds even less optimistic in its on-line document, *Cancer Facts & the War on Cancer:* "If current trends continue, cancer is expected to be the leading cause of death in the United States by the year 2010" [21]. How does this coincide with the goal to "*Eliminate the suffering and death due to cancer by 2015*?" We can only conclude that the NCI is speaking out of both sides of its mouth. On the one hand, it says that cancer is now less of a problem and we will have eliminated the suffering and death by 2015, while on the other hand it says that the problem is growing, presumably to rationalize ever-larger appropriations requests.

THE ANNUAL REPORTS TO THE NATION ON THE STATUS OF CANCER

Since 1998, the American Cancer Society (ACS), the Centers for Disease Control and Prevention (CDC), the NCI, and the North American Association of Central Cancer Registries (NAACCR) have produced an *Annual Report to the Nation on the Status of Cancer*. Like the NCI documents cited above, these reports attempt to convince readers that we are winning the war on cancer, while also sounding alarm bells about the growing impact of cancer in the United States.

The conclusion of the most recent of these reports, issued in September 2003, is exemplary in contradicting information within the *Report* and other NCI documents. We will take a look at its major claims:

> This annual report to the nation suggests that cancer incidence and death rates are leveling off after recent declines. . . . Although considerable progress has been made in reducing the burden of cancer in the U.S. population, a greater effort will be required to meet national health goals, such as the Healthy People 2010 and ACS 2015 challenge goals. Furthermore, as the population of older Americans increases, the number of people diagnosed with cancer is expected to double in the next several decades. Because more of these patients are living longer after a diagnosis of cancer, the strain on cancer control and health care resources to provide treatment and palliation services will increase [22].

The *Report's* first conclusion, ". . . cancer incidence and death rates are leveling off after recent declines" is true—for all cancer sites combined and for certain specific cancer sites. Incidence rates for all cancer sites combined declined from their all-time high of 511 per 100,000 in 1992 to 476 per 100,000 in 1995 and have been hovering around that rate ever since [23]. Mortality rates for all cancer sites combined declined gradually from their all-time high of 216.0 per 100,000 in 1990 to 199.6 per 100,000 in 2000 (a 7.6% decline) [24]. Yet, some racial and sex groups are still experiencing growing incidence and mortality rates for certain cancer sites.

The second conclusion, above, is contradicted by NCI's own data: ". . . considerable progress has been made in reducing the burden of cancer in the U.S. population. . . ." The "burden" of a disease is defined as the number of people living with that disease. The cancer burden has grown constantly in the United States since 1971 (Figure 2). Some of the growth in the cancer burden is due to

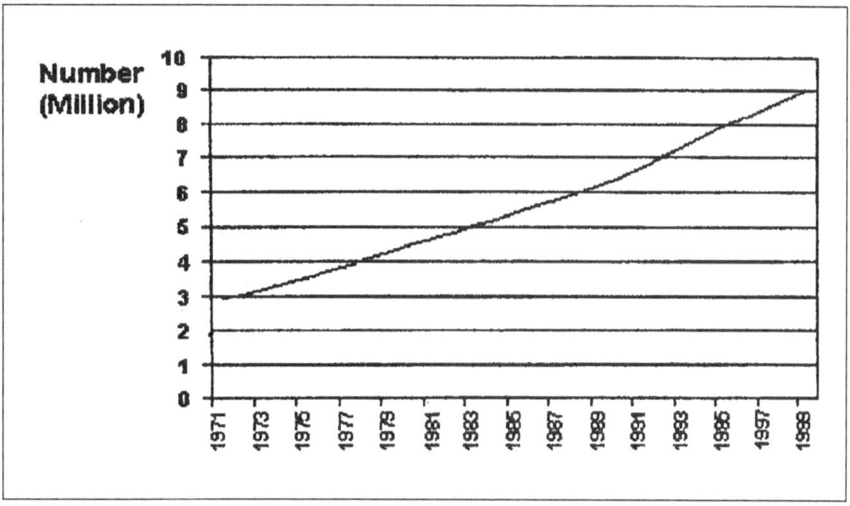

Figure 2. Estimated cumulative number of cancer survivors in the United States for all cancer sites, 1971-1999.
Source: National Cancer Institute, Cancer Control and Population Sciences: Research Findings, http://cancercontrol.cancer.gov/ocs/prevalence/prevalence.html, accessed January 25, 2004.

better treatments or diagnostic services, but it is primarily due to the increase in cancer *incidence* in all age groups and to our aging population, which carries the highest cancer burden.

In 1971, three million Americans were living with cancer or cancer histories. By 2003, that number was nearly 10 million. NCI Director von Eschenbach touts this huge increase in cancer "survivors," rather than "burden," as one of the "increasingly significant dividends" resulting from our "persistence and patience" over the previous 33 years [25]. But, is this a sign of progress? As we have already seen, cancer mortality rates were nearly identical in the late 1940s and 2000 and the rate of new cancer diagnoses steadily increased from 1973, when we first had national incidence data, until the 1990s when they declined slightly. Although incidence for all cancer sites decreased by about 8% from 1992-2000, the number of people living with cancer increased by about 3% [26].

The 2003 *Annual Report's* third conclusion is even more puzzling: "Furthermore, as the population of older Americans increases, the number of people diagnosed with cancer is expected to double in the next several decades." Unfortunately, this may be true, but how does this follow the prior sentence: ". . . considerable progress has been made in reducing the burden of cancer in the U.S. population . . . ?" And, how does this coincide with NCI's goal to "eliminate the suffering and death due to cancer by 2015?" Again, it appears that NCI sounds alarm bells in order to gain approval for its FY 2005 budget proposal of $6.2 billion (a more than $1.4 billion increase over its FY 2004 budget), while proclaiming great strides in the "war on cancer" [27].

The final sentence of the 2003 *Report* also seems to counter NCI's claims of progress: "Because more of these patients are living longer after a diagnosis of cancer, the strain on cancer control and health care resources to provide treatment and palliation services will increase" [28]. Again, we ask: have we won the war yet? Again, it seems that NCI speaks out of both sides of its mouth.

IS BREAST CANCER INFORMATION DISFIGURED?

In addition to cancer in all sites combined, we looked at female breast cancer data because it is the most frequently diagnosed cancer among women other than skin cancers. In the year 2000, 2.2 million women in the United States were living with or had a history of breast cancer [29].

The ACS estimated that 211,300 women would be diagnosed with malignant breast cancer and an additional 55,700 women would be diagnosed with breast cancer *in situ* in 2003. The ACS also estimated that 39,800 women would die of breast cancer in 2003, maintaining breast cancer as the second leading cause of cancer death among women after lung cancer [30].

In 1973, 99 of every 100,000 women were diagnosed with malignant breast cancer and 32 of every 100,000 women died from it. By the year 2000, 135 out of every 100,000 women were diagnosed with malignant breast cancer and 27 of every 100,000 women died from it. The breast cancer mortality rate slowly increased from the 1960 level of 31.7 per 100,000 to 33.2 in 1989 and then gradually declined to 26.8 per 100,000 in 2000 [31].

In the period 1992-2000, breast cancer incidence rose from 131.8 to 135.1 per 100,000, or 2.5%, yet the absolute numbers of women with breast cancer (the "burden") grew by approximately 16%. The breast cancer mortality rate during this period dropped by nearly 16%, but the absolute numbers of deaths declined by only 2.5% [32].

In the period 1998-2000, the lifetime risk for all women of getting malignant breast cancer was one in 7.4. For white women that figure was 1 in 7.0 and for black women it was 1 in 9.9 [33]. The notable exception to this pattern is that black women under 40 are more likely than white women to receive a breast cancer diagnosis. Yet, black women of all age groups face a higher risk of dying of breast cancer than white women. Breast cancer mortality was

15% higher for black women than for white women in 1990. By 2001, that difference had grown to 34% [34] (Figure 3).[3]

MAMMOGRAPHY: SCHMAMMOGRAPHY![4]

The 2003 *Annual Report to the Nation on the Status of Cancer* devotes most of its pages to detailed tables on: the prevalence of smoking, tobacco control, and lung cancer death rates; the prevalence of mammography screening and breast cancer incidence and mortality for black and white women by state; the prevalence of prostate-specific antigen testing and prostate cancer incidence and death rates among black men and white men; and the prevalence of colorectal cancer screening and colorectal cancer incidence and death rates for blacks and whites. It's been well demonstrated that less smoking leads to fewer deaths from lung cancer, but we were curious to find out what promise the attention to screening and mortality holds toward NCI's goal to "eliminate the suffering and death due to cancer by 2015." We decided to take a closer look at the *Report's* data on mammography prevalence and breast cancer.

The *Report* provides data by state on the percentages of women who had a mammogram within the last two years in the year 2000, breast cancer incidence for the period 1996-2000, and breast cancer mortality for the period 1996-2000. For white women, we found a correlation coefficient between mammography prevalence and breast cancer incidence of 0.25, meaning that to a small extent the more mammography, the more breast cancer. For black women, the correlation coefficient of the same data was 0.50—an even stronger association between mammography and breast cancer. For mammography prevalence and breast cancer mortality, we found a correlation coefficient of 0.45 for white women and 0.05 for black women. Again, they are both positive associations, but the latter correlation is weak. We expect mammography to find breast cancers that would not otherwise be found (including those that would not necessarily be life threatening). We also expect the vigorous pressure on women to get mammograms to lower, not raise, breast cancer mortality rates. One caution in interpreting this data on mammography and breast cancer mortality, is that it looks at rates in the same time frame, 1996-2000, and it would be preferable to look at mortality rates somewhat later.

We then wanted to see if we could find evidence in other U.S. data that more mammography lowers breast cancer mortality. The longest lag time we could find for comparison was five years and we restricted the data to women (of all races) age 50 and over, whom mammography is claimed to benefit most. Comparing mammography prevalence data for 1995 with breast cancer mortality

[3] SEER data by race are general available only for whites and blacks.

[4] The title of a 2003 talk by B. Brenner, *Mammography, Schmammography: The Politics of Breast Cancer,* www.bcaction.org.

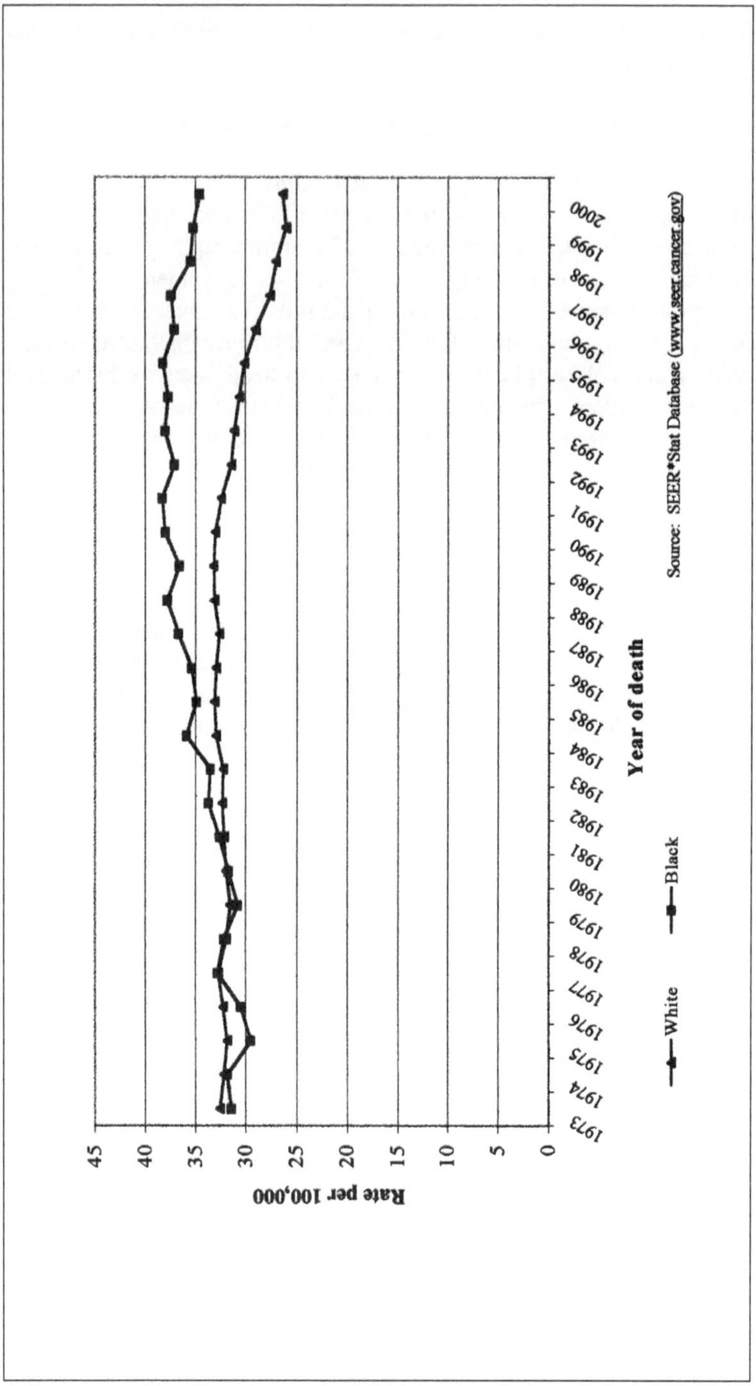

Figure 3. Trends in U.S. female breast cancer mortality rates by race, 1973-2000.
Source: SEER*Stat Database (www.seer.cancer.gov).

data for 2000 by state, we found a correlation coefficient of –0.055, meaning that the screening and mortality data were *not* correlated.

Numerous studies have questioned the value of mammography, notably the meta-analyses by two Swedish researchers published in *The Lancet* in 2000 and 2001 [35]. In addition, the *Journal of the National Cancer Institute* published results of a major study on the Canadian National Breast Screening Study showing that mammography combined with physical breast exams provided no reduction of breast cancer mortality for women in their 50s [36]. On top of this evidence, it's well demonstrated that the ionizing radiation involved in mammography is known to cause breast cancer.

IT'S NOT THAT WE'RE GETTING OLDER

You might say, well of course, if we're not dying of other causes, then we are living long enough to develop and die of cancer, right? Not exactly. All of the data cited above are standardized to the 2000 U.S. population. They reflect the fact that our population is now older and calculate rates for all prior years as if we had the older, year 2000 age distribution throughout the 20th century. The change in standard population from 1970 to 2000 noticeably elevates cancer rates, but trends are unaffected. The aging population and the decline in heart disease death rates do play a role in why cancer death rates have essentially held steady since World War II, but the point is: we are not winning the war on cancer. If we were, we wouldn't have the same cancer mortality rate we had in the late 1940s and we wouldn't have the enormous increases in cancer incidence that occurred during the 20th century.

IT'S NOT THAT THEY CHANGED THE INTERNATIONAL CLASSIFICATION OF DISEASES

In some of its documents, the NCI has asserted that the change from the use of the International Classification of Diseases (ICD) version 9 to ICD version 10 has caused cancer statistics to appear higher than they would have without the change in ICD. The 2002 *Annual Report* is the first in this series to include the change from ICD 9 to 10.

Anderson, Miniño, Hoyert, and Rosenberg described the minimal impact this change should have had on cancer data: "The comparability ratio for Malignant neoplasms . . . is 1.0068. This indicates that the number of deaths due to malignant neoplasms remained stable across revisions." A comparability ratio of 1.0068 indicates that less than 1% of the difference in death rates was due to this change in coding systems. The authors also explained with words that overall cancer data was not affected: "Although these changes do not have an impact on the malignant neoplasm category as a whole, they do have an impact on the distribution of underlying cause of death among various subcategories of malignant neoplasms" [37].

Finally, some deaths were shifted into malignant neoplasms while others were shifted out. A "substantial number of deaths were classified under Malignant neoplasms in ICD-10 that were not Malignant neoplasms in ICD-9." Most of them were pneumonia in ICD-9. Also, many deaths due to malignant neoplasms in ICD-9 were shifted out of malignant neoplasms in ICD-10, usually to categories of HIV or neoplasms of uncertain behavior [37].

Thus, it's not the change in the ICD coding system that keeps cancer death rates from declining.

CONCLUSION

We are far from winning the war on cancer. Mortality rates for all cancer sites combined have been sluggish for the last 60 years and incidence rates for all cancer sites have generally been on the increase for the last 30 years (the period for which we have the best data). The $50 billion spent on the "war on cancer" since 1972 has contributed to modest improvements in cancer treatments and diagnostic techniques, but it does not appear to have eased the cancer burden. The number of people living with cancer grew from three million in 1971 to nearly ten million in 2003.

To make any appreciable advances toward NCI's ambitious goal to "eliminate the suffering and death due to cancer by 2015," we must focus our attention on preventing cancers with demonstrated causes, particularly occupational and environmental causes. We have learned enough about causal factors for many cancers that not acting to remove or reduce known and suspected risk factors is unconscionable. The only acceptable level of carcinogenic exposures is zero. Even if we were to start a broad-sweeping effort now to rid the environment and work place of carcinogens, our efforts would not be evident by 2015. They might be evident by 2025 or 2030.

We have seen that removal of known carcinogens from the workplace and the general environment can reduce cancer rates. One recent example of this among many is Hardell's and Eriksson's research showing that a ban on two pesticides linked to non-Hodgkin lymphoma (NHL) was the likely explanation for the subsequent reduction in NHL rates in Sweden and other countries [38].

Our failures in the "war on cancer" have engendered despair, as well as a multi-billion dollar industry dealing with cancer research, diagnosis, treatment, and marketing products to "look good, feel better." Deborah Forter of the Massachusetts Breast Cancer Coalition wrote:

> We are being lulled into a mindset that winning the war on cancer is learning to live with cancer as a chronic disease. The marketing industry has taken advantage of this epidemic to promote products and garner publicity. The incidence rates continue to increase, women are being diagnosed at

increasingly younger ages, and the promises of bigger and better treatments are proving to be more marketing hype without getting us closer to the end of this epidemic [39].

In the words of Barbara Brenner of Breast Cancer Action, "We don't need more war analogies about cancer. Let's talk about . . . cancer as a human problem instead and start thinking about the human impacts of how we approach the disease" [40].

REFERENCES

1. American Cancer Society, *Cancer Facts & Figures 2003*, p. 2, accessed February 1, 2004 at http://www.cancer.org/docroot/STT/stt_0_2003.asp?sitearea=STT&level=1, 2003.
2. National Center for Health Statistics, National Vital Statistics Reports, Vol. 52, No. 9, November 7, 2003, *Deaths: Leading Causes for 2001, Table 1. Deaths, percentage of total deaths, and death rates for the 10 leading causes of death in selected age groups, by race and sex: United States, 2001,* accessed February 7, 2004 at http://www.cdc.gov/nchs/data/nvsr/nvsr52/nvsr52_09.pdf.
3. Surveillance, Epidemiology, and End Results (SEER), *Cancer Facts and the War on Cancer*, accessed January 25, 2004 at http://training.seer.cancer.gov/module_cancer_disease/unit5_war_on_cancer.html.
4. U.S. Department of Health and Human Services, National Institutes of Health, *The Nation's Investment in Cancer Research, A Plan and Budget Proposal for Fiscal Year 2005*, prepared by the director of the National Cancer Institute, Andrew C. von Eschenbach, p. iii, October 2003, accessed January 19, 2004 at http://plan.cancer.gov/.
5. National Center for Health Statistics, *Health, United States, 2003, Table 29: Age-adjusted death rates for selected causes of death, according to sex, race, and Hispanic origin: United States, selected years 1950-2001,* accessed January 11, 2004 at http://www.cdc.gov/nchs/hus.htm and National Center for Health Statistics, National Vital Statistics System, *Age-adjusted death rates for selected causes by race and sex using year 2000 standard population: Death registration states, 1900-32 and United States, 1933-59,* accessed January 11, 2004 at http://www.cdc.gov/nchs/data/statab/hist293_0059.pdf.
6. Surveillance, Epidemiology, and End Results (SEER), *Cancer Statistics Review 1975-2000*, accessed February 1, 2004 at http://seer.cancer.gov/csr/1975_2000/results_merged/sect_01_overview.pdf.
7. National Center for Health Statistics, *Health, United States, 2003, Table 29: Age-adjusted death rates for selected causes of death, according to sex, race, and Hispanic origin: United States, selected years 1950-2001,* accessed January 11, 2004 at http://www.cdc.gov/nchs/hus.htm and National Center for Health Statistics, National Vital Statistics System, Age-adjusted death rates for selected causes by race and sex using year 2000 standard population: Death registration states, 1900-32 and United States, 1933-59, accessed January 11, 2004 at http://www.cdc.gov/nchs/data/statab/hist293_0059.pdf.
8. National Center for Health Statistics, *Health, United States, 2003, Table 29: Age-adjusted death rates for selected causes of death, according to sex, race, and Hispanic origin: United States, selected years 1950-2001*, accessed January 11, 2004

at http://www.cdc.gov/nchs/hus.htm and National Center for Health Statistics, National Vital Statistics System, *Age-adjusted death rates for selected causes by race and sex using year 2000 standard population: Death registration states, 1900-32 and United States, 1933-59,* accessed January 11, 2004 at http://www.cdc.gov/nchs/data/statab/hist293_0059.pdf.

9. National Center for Health Statistics, National Vital Statistics Reports, Vol. 52, No. 9, November 7, 2003, *Deaths: Leading Causes for 2001, Table C. Deaths and percentage of total deaths for the 10 leading causes of death: United States, 2000-2001,* accessed February 7, 2004 at http://www.cdc.gov/nchs/data/nvsr/nvsr52/nvsr52_09.pdf.
10. National Cancer Institute, *NIH Almanac,* accessed January 24, 2004 at http://www.nih.gov/about/almanac/organization/NCI.htm.
11. U.S. Department of Health and Human Services, National Institutes of Health, *The Nation's Investment in Cancer Research, A Plan and Budget Proposal for Fiscal Year 2005*, prepared by the director of the National Cancer Institute, Andrew C. von Eschenbach, p. vi, October 2003, accessed January 19, 2004 at http://plan.cancer.gov/.
12. U.S. Congress, *The National Cancer Act of 1971,* Public Law 92-218, 92nd Congress, S. 1828, December 23, 1971, Section 3, accessed January 16, 2004 at http://www3.cancer.gov/legis/1971canc.html.
13. S. S. Epstein, Evaluation of the National Cancer Program and Proposed Reforms, *American Journal of Independent Medicine*, Vol. 24, No. 1, pp. 109-33, 1993.
14. Surveillance, Epidemiology, and End Results (SEER), Cancer Query Systems, accessed January 24, 2004 at http://seer.cancer.gov/canques/.
15. National Center for Health Statistics, *Health, United States, 2003, Table 29: Age-adjusted death rates for selected causes of death, according to sex, race, and Hispanic origin: United States, selected years 1950-2001*, accessed January 11, 2004 at http://www.cdc.gov/nchs/hus.htm and Surveillance, Epidemiology, and End Results (SEER), Cancer Query Systems, accessed January 24, 2004 at http://seer.cancer.gov/canques/.
16. Surveillance, Epidemiology, and End Results (SEER), *Cancer Facts and the War on Cancer*, accessed January 25, 2004 at http://training.seer.cancer.gov/module_cancer_disease/unit5_war_on_cancer.html.
17. Cost of the War in Iraq, accessed on February 7, 2004 at http://www.costofwar.com/. The figure includes interest payments on borrowed funds.
18. National Center for Health Statistics, *Health, United States, 2003, Table 38: Death rates for malignant neoplasms, according to sex, race, Hispanic origin, and age: United States, selected years 1950-2001*, accessed January 11, 2004 at http://www.cdc.gov/nchs/hus.htm.
19. U.S. Department of Health and Human Services, National Institutes of Health, *The Nation's Investment in Cancer Research, A Plan and Budget Proposal for Fiscal Year 2005*, prepared by the director of the National Cancer Institute, Andrew C. von Eschenbach, p. iii-iv, October 2003, accessed January 19, 2004 at http://plan.cancer.gov/.
20. U.S. Department of Health and Human Services, National Institutes of Health, *The Nation's Investment in Cancer Research, A Plan and Budget Proposal for Fiscal Year 2005*, prepared by the director of the National Cancer Institute, Andrew C. von Eschenbach, p. xii, October 2003, accessed January 19, 2004 at http://plan.cancer.gov/.

21. Surveillance, Epidemiology, and End Results (SEER), *Cancer Facts and the War on Cancer*, accessed January 25, 2004 at http://training.seer.cancer.gov/module_cancer_disease/unit5_war_on_cancer.html.
22. H. K. Weir, M. J. Thun, B. F. Hankey, et al., Annual Report to the Nation on the Status of Cancer, 1975-2000, Featuring the Uses of Surveillance Data for Cancer Prevention and Control, *Journal of the National Cancer Institute*, Vol. 95, No. 17, September 3, 2003, p. 1296.
23. Surveillance, Epidemiology, and End Results (SEER), Cancer Query Systems, accessed January 24, 2004 at http://seer.cancer.gov/canques/.
24. National Center for Health Statistics, *Health, United States, 2003, Table 29: Age-adjusted death rates for selected causes of death, according to sex, race, and Hispanic origin: United States, selected years 1950-2001*, accessed January 11, 2004 at http://www.cdc.gov/nchs/hus.htm.
25. U.S. Department of Health and Human Services, National Institutes of Health, *The Nation's Investment in Cancer Research, A Plan and Budget Proposal for Fiscal Year 2005*, prepared by the director of the National Cancer Institute, Andrew C. von Eschenbach, p. iii, October 2003, accessed January 19, 2004 at http://plan.cancer.gov/ and Surveillance, Epidemiology, and End Results (SEER), Cancer Query Systems, *US Estimated Complete Prevalence Counts on 1/1/2000*, accessed January 25, 2004 at http://seer.cancer.gov/canques/. See also NCI's graph of estimated numbers of people living with cancer, 1971-1999 at http://cancercontrol.cancer.gov/ocs/prevalence/prevalence.html.
26. Surveillance, Epidemiology, and End Results (SEER), *Cancer Statistics Review 1975-2000*, accessed February 1, 2004 at http://seer.cancer.gov/csr/1975_2000/results_merged/sect_01_overview.pdf.
27. U.S. Department of Health and Human Services, National Institutes of Health, *The Nation's Investment in Cancer Research, A Plan and Budget Proposal for Fiscal Year 2005*, prepared by the director of the National Cancer Institute, Andrew C. von Eschenbach, p. vi, October 2003, accessed January 19, 2004 at http://plan.cancer.gov/.
28. H. K. Weir, M. J. Thun, B. F. Hankey, et al., Annual Report to the Nation on the Status of Cancer, 1975-2000, Featuring the Uses of Surveillance Data for Cancer Prevention and Control, *Journal of the National Cancer Institute*, Vol. 95, No. 17, September 3, 2003, p. 1296.
29. Surveillance, Epidemiology, and End Results (SEER), Cancer Query Systems, accessed January 24, 2004 at http://seer.cancer.gov/canques/.
30. American Cancer Society, *Breast Cancer Facts & Figures 2003-2004,* p. 1, accessed February 1, 2004 at http://www.cancer.org/docroot/STT/stt_0_2003.asp?sitearea=STT&level=1, 2003.
31. National Center for Health Statistics, *Health, United States, 2003, Table 29: Age-adjusted death rates for selected causes of death, according to sex, race, and Hispanic origin: United States, selected years 1950-2001*, accessed January 11, 2004 at http://www.cdc.gov/nchs/hus.htm and Surveillance, Epidemiology, and End Results (SEER), Cancer Query Systems, accessed January 24, 2004 at http://seer.cancer.gov/canques/.
32. Surveillance, Epidemiology, and End Results (SEER), Cancer Query Systems, accessed January 24, 2004 at http://seer.cancer.gov/canques/ and SEER, *Cancer*

Statistics Review 1975-2000, Figure I-17 and I-18, accessed February 1, 2004 at http://seer.cancer.gov/csr/1975_2000/results_merged/sect_01_overview.pdf.

33. Surveillance, Epidemiology, and End Results (SEER), *Cancer Statistics Review 1975-2000*, accessed February 1, 2004 at http://seer.cancer.gov/csr/1975_2000/results_merged/sect_01_overview.pdf.
34. National Center for Health Statistics, *Health, United States, 2003, Highlights*, accessed January 11, 2004 at http://www.cdc.gov/nchs/products/pubs/pubd/hus/highlits.pdf
35. P. C. Gotzsche and O. Olsen, Is Screening for Breast Cancer with Mammography Justifiable?, *The Lancet*, Vol. 355, pp 129-134, 2000 and Cochrane review on screening for breast cancer with mammography, *The Lancet*, Vol. 358, pp 1340-42, 2001.
36. A. B. Miller, T. To, C. J. Baines, and C. Wall, Canadian National Breast Screening Study-2: 13-Year Results of a Randomized Trial in Women Aged 50-59 Years, *Journal of the National Cancer Institute*, Vol. 92, No. 18, September 20, 2000.
37. R. N. Anderson, A. M. Miniño, D. L. Hoyert, and H. M. Rosenberg, Comparability of Cause of Death Between ICD-9 and ICD-10: Preliminary Estimates, CDC, *National Vital Statistics Reports*, Vol. 49, No. 2, May 18, 2001.
38. L. Hardell and M. Eriksson, Is the Decline of the Increasing Incidence of Non-Hodgkin Lymphoma in Sweden and Other Countries a Result of Cancer Preventive Measures?, *Environmental Health Perspectives*, Vol. 111, No. 14, 2003.
39. D. Forter, The Misguided Illusion of Breast Cancer Awareness, *Stop the Epidemic!*, Newsletter of the Massachusetts Breast Cancer Coalition, No. 28, Fall 2003.
40. B. Brenner. From the Executive Director: Wagin' War, Making Connections. Breast Cancer Action *Newsletter #78: September/October 2003,* accessed February 2, 2004 at http://www.bcaction.org.
41. National Center for Health Statistics, *Health, United States, 2003*, *Table 29: Age-adjusted death rates for selected causes of death, according to sex, race, and Hispanic origin: United States, selected years 1950-2001*, accessed January 11, 2004 at http://www.cdc.gov/nchs/hus.htm and National Center for Health Statistics, National Vital Statistics System, *Age-adjusted death rates for selected causes by race and sex using year 2000 standard population: Death registration states, 1900-32 and United States, 1933-59,* accessed January 11, 2004 at http://www.cdc.gov/nchs/data/statab/hist293_0059.pdf.

Note: Originally published in *New Solutions, 14*(2), pp. 109-124, 2004.

Section III

SOLUTIONS SCIENCE

CHAPTER 12

What Is Yet To Be Done

Barry Commoner

The environmental crisis expresses the relation between science and society in a special way: it illustrates the overriding importance of action. Action-oriented decisions—for example, whether to stop global warming for the sake of people or in order to conserve the natural world—profoundly affect the relationship. Both the post-World War II changes in production technology, which gave rise to the environmental crisis, and the failed effort to resolve it by the strategy of control, lead to a common conclusion: Environmental pollution is an incurable disease: it can only be prevented. By design, production technologies must be compatible with environmental quality. This introduces a social interest in what is widely regarded as a private prerogative: the decisions that determine what is produced and by what means. Environmental quality is then an aspect of political economy, requiring, for example, national, democratically determined, industrial and agricultural policies. Such a sweeping transformation of production can be powerfully inspired by a vision of the economic renaissance that would be generated by the new, more productive, technologies. The most meaningful engine of change may be not so much environmental quality as the economic development and growth generated by the effort to improve it.

When, in an excess of politeness, the organizing committee asked me whether I would like to speak at this wonderful symposium, I hesitated—uncharacteristically—because I knew that it would be hard to decide what to talk about. I should have known better, since, for its part, the committee—in keeping with an old Center for the Biology of Natural Systems (CBNS) tradition—did not hesitate to tell me what I ought to say. Perhaps in deference to my now undeniable seniority, the committee's instructions were delivered in a very tactful way, in the form of a four-word code disguised as the title of the symposium: Science and Social Action. Obviously my task was to extract, from only these four words, the committee's detailed instructions on how to make sense of the collective experience, wisdom, courage, and just plain hard work represented by all the good people here today.

In my attempt to break the committee's clever code, I was aided by the fact that the members, thoroughly indoctrinated in CBNS ideology, were likely to think in terms of certain environmental cliches, especially the main one: "Everything is connected to everything else." With that in mind, I quickly understood that the key word in the code was AND, signifying that science *and* society are closely connected, each dependent on the other. Surely I could find a lot to say about this meaty topic. That left the remaining code word: ACTION. This gave me the committee's basic operational instruction: "Don't overdo the philosophizing; just answer the question: 'What is to be done?'"

The interaction between science and social problems and the need to resolve them applies not only to the environment but to a number of other issues that also cry out for action, for example, health and—if we concede that economics is indeed a science—the economy. However, the environmental crisis is special, for it expresses the relation between science and society and the overriding importance of action in a distinctive way that illuminates the wider range of issues as well.

A FUNDAMENTAL FAULT

The environmental crisis arises from a fundamental fault: Our systems of production—in industry, agriculture, energy, and transportation—essential as they are, make people sick and die. As the Surgeon General would say, these processes are hazardous to your health. But that is only the *immediate* problem. Down the line, these same production processes threaten a series of global human catastrophes: higher temperatures; the seas rising to flood many of the world's cities; more frequent severe weather; and dangerous exposure to ultraviolet radiation. The non-human sectors of the living ecosystem are also affected by the crisis: ancient forest reserves are disappearing; wetlands and estuaries are impaired; numerous species are threatened with extinction.

Nevertheless, the environmental crisis is a *human* event; it is caused by what people do, and the ultimate measure of its impact is the health and well-being of people. I start with this assertion because all of us who profess to be environmentalists must decide for ourselves which of two alternative motives justifies environmental action: shall we stop the assault on the Earth's ecosphere for the sake of its human inhabitants—who depend on it—or to conserve the natural world itself? This is an unavoidable choice. For example, it determines whether or not action should be taken to avoid global warming. While, unchecked, the rising global temperature will surely devastate human society, the ecosystems that survive it would be no less "natural" than those of the earlier, equally warm carboniferous era. Only its human impact justifies the effort to prevent global warming.

The modern assault on the environment began about fifty years ago, during and immediately after World War II. I am grateful that my own adult life has covered this span of time, so that I have witnessed most of the notorious environmental blunders that led to the crisis—sometimes simply a bystander, other times an attentive observer, and at least once—in the case of DDT—an unwitting perpetrator. This experience, and my participation in the collective effort to understand and resolve the environmental crisis, has been enlightening. That effort has been marked by major environmental victories and disheartening defeats, and we can learn a good deal from both.

ENTERING THE ARENA

As you have heard, my own entry into the environmental arena was through the issue that so dramatically—and destructively—demonstrates the link between science and social action: nuclear weapons. The weapons were conceived and created by a small band of physicists and chemists; they remain a cataclysmic threat to the whole of human society and the natural environment. World War II had hardly ended when—not satisfied with the wartime bombs that killed hundreds of thousands of people in Japan—the United States and the Soviet Union began testing new and nastier ones, creating enormous amounts of radioactivity that spread through the air worldwide, descending as fallout. Many atomic scientists—alarmed by the consequences of their wartime work—protested. But the tests continued and were even expanded.

The tests were done in secret, marked only by Atomic Energy Commission (AEC) announcements that the emitted radiation was confined to the test area and, in any case, "harmless." This convenient conclusion reflected the AEC's assumption that the radioactive debris would remain aloft in the stratosphere for years, allowing time for much of the radioactivity to decay. But the AEC was wrong; in 1953, shortly after a test explosion, university physicists in Troy, New York, detected significant levels of its radioactivity in rainfall. At about the same time, a then-secret AEC report had concluded that no one was in danger from the radioactive strontium-90 in fallout unless they happened to eat a stray chip of bone in their hamburger. This, too, was wrong: as some of us in St. Louis pointed out, strontium-90 tracks calcium through the ecological food chain, from fallout-contaminated grass, into cows, their milk, and then—because milk, not bone, is their main source of calcium—into children's bones as they grow. Another AEC report concluded that fallout radioactivity levels were too low to "cause a detectable increase in mutations." Again the AEC was wrong; a United Nations scientific committee found that bomb tests would cause up to 100,000 serious genetic defects worldwide.

After 1954 when some of the secret reports were declassified, independent scientists were able to further analyze the fallout data that AEC scientists had developed but had failed to understand. The new analyses confirmed that they had grossly underestimated the dangers: E. B. Lewis, a geneticist at CalTech, showed that iodine 131, a major fallout component, was likely to cause thyroid tumors in children; Linus Pauling, the noted chemist, added carbon 14 to the roster of fallout hazards; Norman Bauer, a chemist at Utah State University, and E. W. Pfeiffer, a University of Montana zoologist, showed that there were high local fallout concentrations near, but outside, the Nevada test site; Erville Graham, a Canadian botanist, showed that the extraordinary capacity of lichens to absorb fallout directly from the air greatly amplified the hazards to native peoples in the Arctic.

FIELD-OF-VISION PROBLEM

The AEC had at its command an army of highly skilled scientists. Although they knew how to design and build nuclear bombs, somehow it escaped their notice that rainfall washes suspended material out of the air; or that children drink milk and concentrate iodine in their growing thyroids. Many people wondered whether the AEC did, in fact, know the truth, but suppressed it in an effort to avoid unfavorable publicity. Of course, this may have occurred. However, I believe that the main reason for the AEC's failure is less complex but equally devastating. The AEC scientists were so narrowly focused on arming the United States for nuclear war that they failed to perceive facts—even widely known ones—that were outside their limited field of vision. As a report from a Pentagon consultant pointed out as late as 1968, the ecological point of view ". . . has been strongly neglected . . . and detailed research is conspicuously absent."

The AEC taught us that when science is forced to serve a powerful self-justified purpose, it becomes too narrow to serve the wider needs of society. It was the independent scientists, outside the AEC, who understood their obligation to society; it was they who met society's need for the truth. But despite their truth-telling and protests, the bomb tests continued and increased in number and intensity.

When the Committee for Nuclear Information (CNI) was organized in St. Louis in 1958, we did something different: we brought scientists *and* civic-minded citizens together. Our task was to explain to the public—in St. Louis and, soon, nationally—how splitting a few pounds of atoms could turn something as mild as milk into a devastating global poison. At about that time, several of us met with Linus Pauling in St. Louis and together drafted the petition, eventually signed by thousands of scientists worldwide, that is credited with persuading President Kennedy to propose the 1963 Nuclear Test Ban Treaty—the first of continuing international actions to fully cage the nuclear beast.

Yet, that did not mean victory, for the U.S. Senate was a nest of cold-warriors and, according to common wisdom, was unlikely to ratify the treaty. But the Senate was besieged by letters, many of them from parents who abhorred the idea of raising their children with radioactive fallout embedded in their bodies.

What convinced the Senators was not so much their constituents' fear of radiation, but that they were *informed;* they knew how to spell strontium-90 and could explain precisely why it was so dangerous. The treaty was easily ratified.

The Nuclear Test Ban Treaty victory dramatized the political power of a public well informed by independent scientists about the technical facts and aroused to action by their own sense of the deadly threat—as close at hand as a glass of milk—of this man-made environmental contaminant. It was as much an environmental victory as a political one. It was an early indication of the collaborative strength of science and social action.

EXTENDING THE MISSION

It was this conclusion that led CNI to become the Committee for *Environmental* Information and extend its mission to the environmental crisis as a whole: the radiation dangers from nuclear power plants; the toxicity of DDT and other pesticides; the hazards of mercury, lead, PCBs and the huge amounts of noxious chemicals that the petrochemical industry produces—and disperses into the environment; the health effects of smog, which accompanied the auto industry's bloated post-war cars; the growing impact of chemical agriculture on water pollution.

These were man-made mistakes that were therefore within our power to remedy. The mistakes were made by the auto companies when they decided to build bigger cars with high compression engines that for the first time emitted nitrogen oxides, which in turn triggered the smog reaction; by the petrochemical industry that persuaded farmers to spread huge amounts of toxic pesticides—many of them carcinogenic—into the environment; by electric utilities that—believing government propaganda that nuclear power would be "too cheap to meter"—built the plants that generate highly radioactive spent fuel, which is yet to be dealt with. The sharp rise in environmental pollution in the twenty years following World War II could be traced to such new technologies of production—new ways of producing electric power, transportation, and food that, while they generated these valuable goods, now violently assaulted the environment as well. The changes were massive and fast: in less than two decades the total amount of automotive horsepower increased four-fold, of inorganic fertilizer nitrogen seven-fold, of synthetic organic chemicals twenty-fold.

In every case, the environmental hazards were made known only by independent scientists who were often bitterly opposed by the corporations responsible for the hazards. It was Rachel Carson, whose brilliant writing made public the toxic impact of DDT on wildlife—for which she was viciously attacked by the chemical industry. Arlie Haagen-Smit, a CalTech plant physiologist, discovered the cause of photochemical smog—a decisive result long denied by the auto industry. Charles Komanoff, among others (including Rob Scott at CBNS), predicted that nuclear power would be uneconomic—a conclusion ridiculed by the industry until it turned out to be so true that no plants have been built in the last twenty years. The occurrence of dioxin in Agent Orange that the U.S. Air Force sprayed on Vietnam was established by a Harvard microbiologist, Matthew Meselson, and its impact on the Vietnamese people documented by Dr. Arnold Schecter of SUNY Binghamton. The persistent work of Dr. Herbert Needleman, Professor of Pediatrics at the University of Pittsburgh, has established the devastating impact of environmental lead on children's intellectual development. Theo Colborn of the World Wildlife Fund and her colleagues have led the way toward an appreciation of the dangerous effects of endocrine disrupters. And I am proud that CBNS has played a role as well: to demonstrate that the huge post-war increase in the use of inorganic nitrogen fertilizer created serious water pollution problems; that large-scale organic agriculture can earn the farmer as much as conventional agriculture; and that—contrary to industry dogma—trash-burning incinerators actually synthesize dioxin as they operate, and are more costly than recycling.

IN APPRENTICES' HANDS

Of course, other scientists and organizations have made their contributions as well. I offer this selected list only to support two important generalizations about the role of science and scientists in generating what we know about the grave environmental impact of the post-war production technologies: First, the scientists, engineers and technologists who designed and built the new technologies—not to speak of their corporate masters—gave no public notice of their environmental faults because they were unaware of them, uninterested or, in some cases, deceitful. The vaunted sorcery of modern technology was hard at work—but environmentally, it was in the hands of apprentices. Second, outsiders were needed to set things right—or at least to help the American people learn what went wrong and why. As individuals, in ad hoc groups, or through more formal organizations, hundreds of scientists, often collaborating with local grassroots organizations, went to church groups, PTAs, and community organizations to explain how fallout, nuclear power, pesticides, and toxic dumps threaten the environment and the health and well-being of the people who live in it. The American people were informed, became concerned, and sought ways to act.

Those of us who have participated in the public debates about environmental issues have often marveled at the public participants' eagerness to learn. Armed with information provided by University of Alaska scientists, Eskimo villagers in Alaska learned enough about the distinctive biological behavior of strontium-90 in the Arctic to defeat the AEC's bizarre proposal to create a new harbor by exploding a hydrogen bomb; antinuclear activists in California became sufficiently acquainted with local geology to pinpoint the danger of an underwater fault near the site of a proposed nuclear power plant.

Earth Day 1970 was irrefutable evidence that the American people understood the environmental threat and wanted action to resolve it. The government quickly responded, and within the year, the National Environmental Policy Act (NEPA) established, as a national purpose, ". . . efforts that will prevent or eliminate damage to the environment." The Environmental Protection Agency (EPA) was created to administer these efforts, and beginning with the Clean Air Act, legislation was quickly enacted to establish specific remedial programs, encompassing the now-massive legislative and regulatory program, which extends into states and municipalities.

Environmental concern is now firmly embedded in public life: in education, medicine, and law; in journalism, literature, and art. It has turned hitherto indifferent politicians into self-proclaimed environmentalists, starting with Richard Nixon—an environmental non-starter who made the issue the centerpiece of his first State of the Union address. Looking back on these changes—or perhaps startled by the latest advertisement of an oil company that has turned itself green—we might be justified in proclaiming victory. Certainly, we have made things happen. But what has motivated environmentalism and, in my view, defines its purpose is the state of the environment itself.

FAR FROM VICTORY

By that measure we are far from victory: Neither the general aim stated in NEPA, nor the specific improvements mandated in the enabling legislation, have come even close to being achieved. The numerical evidence on the required improvements in air quality—which called for 90 percent reduction in pollution within seven years of 1970—is a persuasive example. According to the latest EPA assessment, after twenty-five years the best percentage improvement in emissions of the standard air pollutants (for sulfur dioxide) since 1970 is only 30 percent. Nitrogen oxide emissions have not improved at all over that period. Worse, in almost all cases whatever improvement did occur came to a halt after 1980; since then, except for a slow reduction in carbon monoxide emissions, the curves are flat. And EPA foresees no further improvement; its latest projections of air emissions show slight *increases* for all the standard pollutants from now to 2010, except for a small decrease in sulfur dioxide.

These numbers tell us that the methods that EPA introduced after 1970 to reduce air pollutant emissions worked for a while, but over time have become progressively less effective—and from the 1980s on no longer capable of reducing emissions at all.

The chief remedial method has been the installation of emission control systems—devices attached to the pollutant-generating source (such as autos, power plants, and incinerators) that trap and destroy the pollutants before they enter the environment. The fault is not that the control devices have themselves become less efficient since the 1980s. Rather, a countervailing process has overcome their emission-reducing capability. That process is economic growth: year by year, there are more cars and trucks on the road and more energy generated. As long as a control device is not perfect—that is, does not reduce emissions to zero—this increased activity counteracts the device's ability to reduce environmental pollution, and economic growth becomes the enemy of environmental quality.

This antagonism between the economy and the environment is built into the nature of a control device: for thermodynamic reasons, it becomes progressively more difficult—and therefore more costly—to remove the pollutant as the percent removed rises. For example, controls that remove 70 percent of sulfur dioxide from a coal-burning power plant's flue gas cost $50 per kilowatt of plant capacity; a system that removes 90 percent costs $2,000 per kilowatt, and it would take $4,000 per kilowatt to reach 99 percent—at which point the control system would cost ten times the cost of the power plant itself. As a result, it is simply economically impossible to require controls that even approach zero emissions. In turn, this economic limitation renders the control system vulnerable to the countervailing effect of increased economic activity. By adopting the control strategy, the nation's environmental program has *created* a built-in antagonism between environmental quality and economic growth.

THE FATAL EMBRACE

Tragically, this conflict—as well as the accompanying failure to meet the legislative goals of environmental improvement—could have been avoided if the enabling legislation had required EPA to abide by NEPA's stated purpose to *prevent and eliminate* pollution. By any interpretation, this requirement means *zero* emissions, which, if accomplished, would meet the mandated goals and undo the fatal embrace between the environment and the economy.

Ironically, hidden in the otherwise dismal data on air pollution emission trends, we can find concrete evidence that the strategy of prevention can actually achieve this astounding result. In 1970 U.S. vehicular transportation emitted 180,000 tons of lead into the air; by 1994, emissions had decreased to 1,600 tons—by 99 percent. This was achieved while vehicular transportation—a major economic activity increased—by 50 percent, as measured by fuel consumption. Environmental quality was drastically improved while economic activity grew by the simple expedient of removing lead from gasoline—which *prevented* it from entering the environment. There are a few similar examples of success by prevention in the environmental data. Environmental levels of DDT have decreased sharply—because its use has been banned; in some rivers, phosphate levels have declined a great deal—because phosphate-containing detergents have been banned. Meanwhile, neither agriculture nor detergent production has suffered.

These only-too-rare miracles have been accomplished by a well-known industrial practice: the technology of production—of gasoline, detergents, and cotton (the chief agricultural use of DDT)—has been altered, albeit at the behest of the government.

The task, then, is to apply the principle of pollution prevention to the major production processes that, in their present form, generate the mass of environmental pollution. There are existing pollution-free alternatives to the production technologies that brought on the post-war environmental crisis. The major source of photochemical smog—petroleum-fueled vehicles—can be replaced by emission-free electric vehicles. In turn, power plants now fueled by oil, natural gas, or uranium can be replaced by zero-emission photovoltaic cells or wind generators. The pollution-free alternative to current agricultural practice—which is heavily based on inorganic nitrogen and synthetic pesticides, both major causes of environmental pollution—is organic farming. In specific process industries, for example, paper mills, closed-loop systems that produce no effluents or emissions are feasible—one is already operating in Finland.

The new production technologies may be more economical than the ones they replace. For example, a recent CBNS study shows that the impact of trash-burning incinerators in the states adjacent to the Great Lakes on the airborne dioxin deposited in the lakes can be reduced to zero by diverting the trash to intensive recycling programs. The net economic effect would be a $500-million reduction in annual disposal costs, including the cost of paying off the incinerators' existing debt. As noted earlier, we have also found that large-scale organic agriculture is competitive with conventional practice in the Midwest.

WHY NOT THIS COURSE?

Why hasn't EPA pursued this course? Until 1989, pollution prevention was entirely absent from the EPA program, despite NEPA's assertion that it was the purpose of the national environmental effort. Then in January of that year, Lee M. Thomas, the retiring Administrator of EPA, published a "Pollution Prevention Policy Statement" in the Federal Register declaring that control measures had failed to satisfactorily improve the environment and that only prevention can succeed.

Although Mr. Thomas' departing testament sought to introduce pollution prevention as the EPA's guiding *policy* (that word, after all, appears in the title) in preference to pollution control, in practice EPA has reduced it to a subsidiary, rather than guiding, feature of the national program. Perhaps most indicative of its fate is that Mr. Thomas' portentous Pollution Prevention Policy has been given a cute acronym: P2. This even eliminates the word *policy* from the official acronym, which of course ought to be: P3. Meanwhile, where policy is actually expressed in EPA operations—in the promulgation of regulatory measures—pollution control holds sway. For example, in the latest regulatory document on dioxin emissions (for hazardous waste incineration), EPA prescribes a Maximum Achievable Control Technology.

Unlike EPA, direct public pressure, often organized by grassroots environmental and community organizations, *has* accomplished major pollution prevention measures. DDT fell victim to Rachel Carson's widely read *Silent Spring;* PCBs were banned in response to public outcry over a flurry of accidental releases. More recent examples include: the abandonment by McDonald's of plastic ware largely as a result of a children's campaign ("McToxic") organized by the Citizens Clearinghouse for Hazardous Waste; in many cities, trash-burning incinerator projects have been abandoned in favor of recycling under pressure from local grassroots groups; a consortium of such groups is waging a vigorous campaign to replace hospital waste incinerators with dioxin-free autoclaves.

These efforts are enormously important, for they provide concrete examples of how the abstract idea of "transforming the technology of production" can be turned into reality. But in keeping with the organizing committee's encoded admonition, I must add that there is a great deal yet to be done, beyond these local grassroots victories.

What is needed is a transformation of the major systems of production more profound than even the sweeping post-World War II changes in production technology. Restoring environmental quality means substituting solar sources of energy for fossil and nuclear fuels; substituting electric motors for the internal combustion engine; substituting organic farming for chemical agriculture; expanding the use of durable, renewable, and recyclable materials—metals, glass, wood, paper—in place of the petrochemical products that have massively displaced them.

RELATING ACTION, CORPORATE DECISIONS

In the U.S. economy, the decisions that determine what is produced and by what means are in private, generally corporate, hands. How can the demand for action to improve the quality of the environment, which is deeply embedded in society as a whole, be brought to bear on these private, corporate decisions?

I believe that the first step is to extend the environmental issue into the relevant social, economic, and political arenas. Consider, for example, the decision to replace conventional cars and light trucks with electric vehicles—powered, ultimately, from solar sources. The relevant corporations are reluctant to make this change because, compared with conventional ones, electric vehicles would initially be more costly and more restricted in their uses. Such a shift would damage a corporation's economic interests, they argue, in comparison with firms that refrained from making the change. However, this issue can be dealt with by establishing, as a national *industrial policy,* that all suitable vehicles are to be powered by electricity, placing all of the auto industry's firms on the same level playing field, economically.

There is nothing new about national policies on major social interests such as education or labor—or, for that matter, the environment. After all, despite the economic advantage to firms that employed child labor, it was in the *social* interest, as a national policy, to abolish it—removing that advantage for *all* firms. What is new is that environmentalism intensely illuminates the need to confront the corporate domain at its most powerful and guarded point—the exclusive right to govern the systems of production.

How can the environmental movement come to grips with this deeply rooted privilege that has firmly resisted even public discussion, let alone a proposed diminution? A useful way to approach this question is to think about it directly in economic, rather than environmental, terms. Seen that way, the wholesale transformation of production technologies that is mandated by pollution prevention creates a new surge of economic development. But this would touch on other social concerns as well. The wave of new productive enterprises would provide opportunities to remedy the unjust distribution of environmental hazards among economic classes and racial and ethnic communities. For labor unions, it would represent a source of new jobs and opportunities to advance the cause of a healthy work environment and worker retraining. Indeed, the transformation, although environmentally mandated, may be much more powerfully inspired by the vision of an economic renaissance that would be generated by the new more productive technologies. The most meaningful engine of change, powerful enough to confront corporate power, may be not so much environmental quality, as the economic development and growth associated with the effort to improve it.

A PREVAILING MYTH

Why should environmental advocates be in favor of economic growth, when this seems to fly in the face of the argument, often advanced by some environmentalists, that high rates of production and consumption are the chief *cause* of environmental degradation? That view is based on the assumption that production is *necessarily* accompanied by pollution, so that these two processes rise and fall together. It reflects a prevailing myth that production technology—the high-compression engine, the nuclear reactor, or genetic engineering—is simply the practical application of scientific knowledge and is therefore no more amenable to human judgment or social interests than the laws of thermodynamics, atomic structure, or biological inheritance. The environmental experience has shattered this myth. The high-compression engine and the nuclear reactor were built in response to *human* decisions and their linkage to smog and radioactive waste can be readily broken by building electric vehicles and photovoltaic cells instead.

On the other hand, there are powerful reasons why environmental advocates should favor economic development and growth—*that is based on ecologically benign technologies of production.* The most cogent reason is that the massive transformation of our major systems of production—that is essential to environmental quality—cannot achieve this goal if it is pursued only in developed countries.

The environmental crisis is a global problem and only global action will resolve it. Even concerted action by the northern industrialized countries—where most of the assault on the ecosphere now originates—will not be enough. What is done in developing countries is crucial as well. As the world population rises—90 percent of the increase in developing countries—worldwide production levels will need to increase sharply in order to sustain global economic development. Unless the expanded production facilities are ecologically sound, this process will further degrade the environment.

There are serious constraints on developing countries that, if unrelieved, will greatly reduce their ability to participate in the transition to ecologically sound systems of production. Since for some time the required production facilities—for example, solar energy equipment—would need to be imported, developing countries are potentially a huge market for the new environmentally benign products. In the United States and other developed countries, this demand would hasten the development of the transition and facilitate the growth of the new production facilities.

THE POWERFUL ENEMY

Of course, none of this can happen if we accept the false idea that environmental quality cannot tolerate economic growth. If, as I believe, the purpose of the environmental effort is to improve the health and well-being of people, then we must recognize that the most powerful enemy of human welfare is poverty. And we must remember that the human inhabitants of the Earth's ecosphere are engulfed in a global epidemic of poverty, hunger, and despair. The grim statistics can be summarized in a simple image. As the earth spins through space, a view from above the North Pole would encompass most of the wealth of the world—most of its food, productive machines, doctors, engineers, and teachers. A view from the opposite pole would encompass most of the world's poor. The planet is split by a chasm that separates the North from the South, the rich from the poor.

This global chasm *must* be bridged. This is the rational, logical outcome of the environmental experience. But I say to you that if environmentalism is to be devoted to human welfare, there are reasons more powerful than the environmental ones. Simple morality dictates that the rich should share their productive capacity with the poor. And an even more compelling imperative is justice, for the poor half of the planet has been brought to that plight through the exploitation of its resources and its people by the imperial nations of the North. We, who are environmental advocates, must find a way—for the sake of the planet and the people who live on it—to join a historic mission to end poverty wherever it exists. That is what is yet to be done.

Note: Originally published in *New Solutions, 8*(1), pp. 75-87, 1998.

CHAPTER 13

Good Practice Guidelines for Occupational Health Research Funded by the Private Sector

Margaret Quinn, Charles Levenstein, and Gregory F. DeLaurier

The role of the private sector in funding academic research is increasing and the well-developed guidelines for government-sponsored research do not apply to the academic-industry arena. Good Practice Guidelines for privately funded occupational health research are necessary. Industry sponsors and academic researchers belong to differing systems with differing goals and means to achieve and evaluate them. As a result, problems are inherent in the relationship. Guidelines would benefit industry by providing criteria against which industry-funded research could be judged and evaluated. Guidelines would help university researchers assure that their work is examined and criticized on its merits. Such protection would foster quality research over the long term. Here, we consider the issues involved and explore questions that came out of a workshop convened under the auspices of the Centers for Disease Control and Prevention. When private funding is involved, university scientists have a professional responsibility to ensure that the integrity of researchers, subjects, and the research process is well protected. A body of experience allows us to create Good Practice Guidelines beneficial to all parties.

The purpose of this chapter is to identify essential issues for the development of a practical set of guidelines that may be used in the design and conduct of occupational safety and health research funded by the private sector. Here, the "private sector" is defined primarily as corporations and trade associations, the most likely sources of occupational health research funds outside of the federal government. In addition, the secondary, but significant, role of labor unions in funding occupational health research is briefly considered. Though private foundations and litigation-related service are also sources of funding, their relationship to the conduct of occupational health research is beyond the scope of this work.

The development of a set of guidelines for privately funded occupational health research is necessary for two basic and related reasons. First, the role of the private sector in funding academic research is increasing, and academic institutions are strongly encouraging academic-industry partnerships. There are well-developed guidelines for government-sponsored research, however, these do not apply to the academic-industry arena. Government-funded research procedures that serve to protect its integrity include peer review of the initial research proposal by a government-sponsored and -managed scientific review committee; protocols for the protection of human subjects, including confidentiality, oversight of the conduct of the research by scientific and administrative project officers, and incentives to publish the research findings in the peer-reviewed literature.

Second, while this private sector-academic research relationship offers many benefits, it is also fraught with difficulties, difficulties often described in situational terms—i.e., difficult personalities, corporations, unions, universities—but which are actually rooted in more fundamental if latent problems of differing goals and expectations on both sides of the relationship. Because these problems are indeed so fundamental, they should not be thought of as extrinsic to the research process but, rather, integral to it. Hence the need for the development of "Good Practice Guidelines" that might identify and anticipate, perhaps even alleviate, conflicts between academic researchers and private funders before they occur. Currently, researchers must negotiate the terms of each industry agreement without access to the prior experience of other researchers. Such guidelines could eliminate the need, so to speak, to re-invent the wheel with every new study, saving both parties considerable time and effort.

I. THE NEED FOR GUIDELINES

During the presidencies of Reagan and George H. W. Bush, the private sector was explicitly encouraged to assume the funding of projects that addressed public concerns. In the field of occupational safety and health, this came to mean an increase in industry-academic cooperation and partnership, including an expansion of industry-funded research at universities. These trends continued under President Clinton, and are likely to be continued, if not accelerated, by the administration of George W. Bush.

There are numerous benefits to industry-funded research in occupational health. Industry-academic partnerships reduce government expenditures and foster practical and commercial applications of research that can lead to sustained economic growth. Industry sponsorship can greatly benefit an occupational health study by engaging the full cooperation of an industry in providing plant access and on-site collection of industrial hygiene and epidemiologic data.

Industry health and safety professionals can also be a valuable resource in the conduct of such studies.

At the same time, however, the objectives of industrial sponsors may conflict with those of academic researchers, as these two groups are each part of differing systems with differing goals and means to achieve and evaluate them. The position of the researcher can be succinctly summarized: scientific research involves—by definition—the discovery of new knowledge related to a hypothesis ideally formulated by the researcher, and publication of the results for review and criticism by the larger scientific community. The position of the industrial sponsor is somewhat more complex and can depend on the circumstances.

In most cases, a corporate funder will wish to know if a product is a hazard to the producer, the consumer, or, more generally, the environment. Yet it must also be concerned with its profitability and long-term stability, and thus be in a position to control the information that affects these goals. The conflict arising from the different systems—scientific versus economic—becomes evident when occupational health research findings have potentially negative economic or public relations consequences for the funder. The goal of science, then, may not necessarily serve the interests of corporations—and at times labor unions—when corporations have little control over the findings or the process by which these findings are made public.

Indeed, corporations may view the generation of such knowledge as a liability. The discovery of new health hazards in an industry, for example, may severely jeopardize a corporation's future. It is in the corporation's best interest to know about such a hazard, but a corporate sponsor will be unlikely to agree if it cannot control how, when, and even if such new knowledge becomes public.

The potential liability for the hazard falls squarely within the purview of corporate lawyers, whose job it is to protect the economic and reputational assets of the firm. University researchers and their professional corporate colleagues may find themselves pressured by corporate counsel who wish to see the conduct, content, and release of findings of research closely managed. Corporations often prefer to enter into, or view, their arrangements with university researchers as "contracts" rather than "grants." Contracts are precise legal documents that have pre-determined protocols, with very limited allowance for deviations, and a well-defined "deliverable." All of this means that significant limits may be imposed on the research process.

By necessity, researchers will emphasize the "grant" nature of a funding agreement, that is, view it as an accepted proposal to investigate a hypothesis or set of hypotheses using open-ended methods that may be developed as each step is performed and used to inform the next. Scientific research usually involves more flexibility than contracts allow, with researchers left free to pick up more promising leads. It is thus not possible within scientific research to define each and every technique and procedure a priori, nor is it possible to define clearly the final product except in the most general terms, i.e., a report of the findings, whatever they may be.

At present, some of these issues and anticipated conflicts between researcher and corporate sponsor are negotiated by industry representatives, university researchers (and their offices of sponsored projects), and sometimes by ad hoc scientific advisory boards established for individual projects. For researchers and industry representatives who are both eager to begin work, such negotiations are often seen as peripheral to the proposed study, even though the details of the agreement may, to a significant extent, determine the study's conduct, progress, and ultimate delivery of the findings to the broader scientific community. Only later, while the study is ongoing, will conflicts and problems be discovered. This is a hard way to learn, and perhaps unnecessary if, as we have suggested, good practice guidelines are in place.

Good practice guidelines would benefit industry by providing criteria against which industry-funded research could be judged and evaluated in the scientific literature. As it stands today, any industry-funded study is received with skepticism and often regarded as fatally biased by a large segment of the scientific community, as well as by the general public. With more and more studies being conducted under the sponsorship of private industry, there is a critical need for criteria that can be used to evaluate their scientific validity and legitimacy. This is crucial for the achievement of the ultimate goal of protecting worker health and the formation of sound public policy such protection demands.

Workers and community members who constitute study populations for industry-funded projects would also benefit from guidelines that address issues of confidentiality, informed consent, and communication of study results to study participants. An important aim of good practice guidelines, then, is to provide a basis for workers and labor and community representatives to evaluate a study.

Such guidelines would also help university researchers. Often unable to secure public funding of their research activities and pressured by their universities to cooperate in establishing links with the private sector, researchers in occupational health must turn to industry for funding. These university scientists need to protect their scientific and personal integrity in this new funding context, and good practice guidelines would assist them in assuring that their work is examined and criticized on its merits. Such protection would also go a long way in assuring the development of quality research over the long-term, a process that underlies the development of scientific understanding.

II. ISSUES

Through exploratory discussions with occupational health researchers, and examination of our own experience, we have developed a set of questions and issues that we believe are basic to the ethical conduct of occupational health research funded by the private sector. The issues involve concerns about (A) the independence of scientific investigators, (B) the role of scientific advisory boards, and (C) the rights of workers in the research process and dissemination of findings.

A. Researcher Independence and the Research Contract

Independence and objectivity are central to the conduct of good science and to the functioning of an academic system. However, such independence can run contrary to the dictates of the business system; the notion of researcher independence can conflict with industry goals which may be better served by controlling or managing the research process, product, and timetable. While the researcher is not an employee of the private funder, conflicts may arise so that the industry feels compelled to manage the researcher. Especially for larger studies, flexibility may well be needed on both the part of the researcher and the industry to meet unexpected demands during the course of the research, but a line must be drawn between flexibility and inappropriate interference. Clearly, it is not possible to foresee all problems and completely alleviate the potential for inappropriate control of the research process by the funder. Nevertheless, an important question related to researcher independence is, "To what extent can guidelines address inappropriate interference by a funder without becoming so cumbersome as to interfere with the research process itself?"

Researcher independence is intended to be addressed in part by the university-funder contract. Some academic institutions have standard contracts for industry-sponsored research. These lay out general university policy concerning protection of human subjects, rights to alter research protocols, interference in the conduct of research, and regulations concerning publication. Even in these cases, though, the terms of research do not address the full range of issues that relate to researcher independence. In addition, the terms that do exist are usually negotiable by the funder.

As it may be crucial to the future independence of a researcher during her or his study, a number of key questions should be raised about the contract. These questions have been derived from our own and other occupational health researchers' past experience with contractual issues as well as from a more general concern with researcher independence.

1. *Ownership and use of the data collected in the study.* Who owns the data, including all material such as job records assembled by the researcher, air sampling results, and the like? Who has a right to a copy of those data? Does the researcher have the right to use the data during or after the close of the study for the submission of new grant proposals?

2. *Confidentiality.* Are medical and exposure data collected by the researcher concerning the workers kept confidential from the sponsor? For instance, is the employer provided documentation that a worker has been exposed to a substance

for a specific length of time? Are exposure data with personal identifiers given to the company?

3. *Researcher independence.* Does the researcher have the right to alter protocols in light of new research developments? For example, if the investigation incidentally turns up a hazardous substance not the subject of the original study protocol but perhaps related to the study outcome, may the researcher pursue this new line of investigation? Does the sponsor have the right to approve protocols or to give the final review of protocols?

4. *Public dissemination of study methods and findings and publication rights.* Does the researcher have the right to make public data collected during the study and the findings of the research in the form of publications and oral presentations? If so, does the sponsor have the right to review the proposed publication? What is the amount of review time required by the sponsor, if a review period is necessary at all? Must the researcher, say, submit abstracts for an oral presentation for sponsor review? Should the sponsor have the right to review all slides and overheads prior to presentation? If so, what specifically is the period of time required for this review?

5. *Dispute resolution procedures.* How are disagreements or disputes over scientific questions to be resolved? How are disputes over the interpretation of results to be resolved? Are there institutional protections, within the university or elsewhere, for intellectual property and mechanisms for the resolution of disputes among authors? For government-funded studies, there are guidelines concerning scientific misconduct. These address issues such as the falsification of data, plagiarism, etc. It is by no means clear that such guidelines apply to research funded by the private sector.

Overall, there is considerable variation in the level of detail among university-industry-sponsored contracts and in the specific provisions regarding researcher independence. Contracts will vary, for example, regarding the amount of time a sponsor is given to review written and oral presentations of a study's methods and findings. The typical range is 30 to 60 days, although an unspecified amount of time may be allowed.

These funder review-time provisions can have significant practical implications for researchers and their ability to present data publicly. Some conferences do not announce the call for abstracts much more than 60 days before the abstract submission deadline. This makes it difficult for researchers to present work requiring a long funder review period. It may also not be feasible for researchers to have all slides or overheads prepared for a conference 30 to 60 days in advance.

All university contracts, then, are not the same. Researchers should not assume that the knowledge of one university contract will apply to another. Contract variation could affect researchers when moving to a new university or perhaps even when subcontracting industry-sponsored research from another university.

Provisions of standard industry-sponsored research contracts are also, it appears, negotiable by the funder and the university, and the researcher would do well to understand the implications of certain changes in contract provisions. Good practice guidelines might assist researchers in such contract negotiation. The extent to which guidelines could address situations of potential conflict that are not usually covered by university research agreements is thus an important area of investigation.

B. The Role of Scientific Advisory Boards

Scientific advisory boards may serve several purposes concerning the relationship between academic scientists and industry. Certainly, they are created to provide scientific expertise in occupational health and to give advice to industry or industry-union funders. They may also provide legitimacy to industry efforts to deal with health and safety problems. Committees may be ad hoc, designed to deal with particular problems that have arisen, or they may be permanent and have the broad role of monitoring worker health in the industry. They may be the creation of one company or of an industry association, and may or may not have been created with relevant labor union partners. Few boards seem to have explicitly stated criteria for membership. Leading academic scientists, at times scientists from government or independent research institutes, and scientists in the employ of the interested parties—management, labor, or both—may be found on advisory boards. Generally, the academic scientists are paid as consultants by the sponsor. There is the potential for conflicts of interest to arise, however, if members of a board seek funding from the sponsors for their own studies while they are at the same time judging the research proposals or findings of others.

Conflicts of interest also depend on the role of a particular advisory board within the research process. First, a committee may be involved in defining a problem or problems, or it may be presented with an issue by the sponsor. In either case, it is important that the committee has full information, including access to industry data and internal reports. Second, the committee may be mandated to write a "request for proposals" based on the defined problem. This is quite distinct from a third function, administering a competition among researchers.

This third function may include judging the qualifications of competitors, reviewing their proposals, and even negotiating the actual contract. Sometimes the committee will simply review the proposals and make recommendations to the company or industry association, and the final negotiations will be left to the sponsor without further involvement of the outside advisors. Occasionally the committee may serve as a peer review group, receiving and commenting on research protocols, methods, and interim reports. Scientific advisory boards may also provide peer review of findings and their interpretation. The sponsor may even empower the committee to consider and develop rules concerning the publication of results.

Privately constituted scientific advisory boards may play a key role in legitimizing research funded by the private sector. The conditions under which members are selected and the authority and degree of autonomy with which they operate are thus important matters of public policy and should be considered in the construction of any good practice guidelines. Certainly the rules of committees matter greatly to their own members and to the researchers about whose work they advise.

C. The Rights of Workers in Occupational Health Research

Concern about the rights of workers in occupational health research, as well as in company-sponsored screening and surveillance programs, has been voiced in the relevant literature and discussed at professional meetings. Traditionally, these discussions have focused on issues of informed consent, confidentiality, and workers' rights to study results, especially for workers determined to be at high risk of occupational disease and injury. More recently, models of participatory research have expanded the discussion of worker rights to include all components of the research process.

Issues of worker rights in occupational health research, then, are not simply matters of ethics and politics, but can affect the conduct and end results of research directly. In many, but not all, cases, occupational health researchers rely on the cooperation of their worker/subjects when collecting both exposure and health effects information. Unless informed of the nature and intent of the study, subjects may express their "consent" in ways that may jeopardize the validity of the study results. Workers not informed about the conduct and purpose of a study may, for example, alter their work patterns when they see industrial hygienists observing their job tasks, mistaking the hygienists for industrial engineers doing time-motion studies. In more extreme cases, uninformed workers may be resistant to or even sabotage the collection of data.

Moreover, worker/subjects are in a considerably different position from patients or other subjects who usually, but not always, volunteer their participation in clinical trials or community-based epidemiologic research. Worker/subjects are dependent on their employers for their livelihoods, and thus can be constrained when making choices about participation in employer-supported research projects.

Issues surrounding worker rights must be addressed when developing good practice guidelines. This is so not only because of a moral duty to research subjects or a political agenda that advocates a fairer allocation of "control" in the workplace, but because research results will invariably be called into question if worker/subjects have not been willing participants in the study.

III. WORKSHOP QUESTIONS FOR THE DEVELOPMENT OF GOOD PRACTICE GUIDELINES

On November 1, 1996 a workshop funded by the Centers for Disease Control and Prevention (CDC) was held on the campus of the University of Massachusetts Lowell. Its purpose was to investigate further the possibility of developing good practice guidelines for occupational health research funded by the private sector. Participants were primarily occupational health researchers from academia.

The majority of discussion took place within three small groups of approximately ten people each, organized by the topics presented above: researcher independence, scientific advisory boards, and worker rights and notification. To facilitate their work, each small group was given a set of questions related to three key issues for each topic and were then asked to formulate draft guidelines around these questions. The questions for the three groups (in places, slightly modified from their original form for clarity) are as follows.

A. Researcher Independence

1. Study conduct:

- How should the need for deviations from the original study protocol, deemed necessary due to new information discovered during the course of the research, be addressed?
- How should disputes about deviation be resolved?

2. Use of data, mid-study and at the close of the study:

- Can a researcher use the study protocols and data collected during the course of a study to write new research grant proposals, including those to potential funders other than the industry in question?
- If so, should the researcher be required to notify the industry funder?
- Should the industry have the right to review the new grants?
- If so, what time periods should be involved with the review?

3. Public dissemination and publication of data and study findings:

- Who owns the data collected during the study?
- Who should be permitted to have copies of the data or portions of the data for further study?
- Does the researcher have the right to publish data or protocols during the conduct of the study?
- Does the researcher have the right to do so at the close of the study?
- If so, does the funder have the right to review written and oral presentations?
- If so, what time period should be granted for such review?

B. Scientific Advisory Boards

1. *Independence:*

- How important is it that a scientific advisory committee be independent of the funder?
- How can the appropriate degree of independence be assured?

2. *Role with respect to researchers:*

- Should the scientific advisory board play a continuing role in the conduct of the study?
- Should the scientific advisory board serve as a "buffer" between industrial funders and researchers?

3. *Role in interpretation and dissemination of results:*

- Should the scientific advisory board play a role in the interpretation of studies?
- Does the scientific advisory board have responsibilities concerning the dissemination of research results?

C. Worker Rights and Notification

1. *Informed consent:*

- What elements of "informed consent" should be included in the conduct of occupational health research?
- How should investigators ensure that these elements have been addressed effectively?
- Are there situations or types of studies in which informed consent of workers is not required or needed?
- How can these situations be correctly identified?

2. *Notification and dissemination of findings:*

- How and when should the research findings be presented or reported to workers and individual study subjects?
- What is the role of the researcher in this process?
- What should be included and addressed when results are reported to workers? For example, should the presentation of study results include a discussion of possible changes or interventions that might be needed in the workplace?
- What is the obligation of researchers to report study results to other groups, e.g., other workers at risk, health care providers, regulatory agencies?

3. *Confidentiality:*

- What data or information should be considered confidential—unavailable in any form to study sponsors and company personnel—in occupational health research studies?
- How can the confidentiality of these data be ensured?

The discussions that ensued during the workshop were both lively and, at times, heated. Overall, attitudes among participants towards the formulation of good practice guidelines ran the gamut from the feeling that such guidelines were not needed to the position that under no circumstances should occupational health researchers accept funding from corporations or trade associations. Also, while many worthwhile suggestions were made for effective guidelines, participants overall felt these suggestions were still too preliminary to be made public. Finally, it was felt that the work of formulating guidelines should be more representative of the occupational health profession as a whole, and thus also include input from researchers working for industry, labor unions, and government, not only from academia.

IV. CONCLUSION: WHERE WE GO FROM HERE

The funding of occupational health research by the private sector—while, as we have noted, beneficial in many ways—is indeed problematic. Still, trends point toward ever increasing funding from this source. Given this reality, good practice guidelines are essential. The National Institute for Occupational Safety and Health should have a strong role in ensuring responsible conduct of occupational health research, even though the agency does not fund all studies that merit support. Having promulgated a National Occupational Health Research Agenda (NORA), the agency has a special duty to assure that research used for public policy purposes meets commonly understood standards for good practices. As national policy has encouraged the private sector funding of research, it is appropriate that national policy be developed to help guide its conduct.

We have outlined in this chapter why there is such a need and suggested specific topics and questions that might be included in these guidelines. But what we have presented is just a beginning. As the workshop experience showed, much needs to be discussed and debated. This chapter is an invitation to do just that. As occupational health investigators, we, as a profession, have a responsibility to insure that when private funding is involved, the integrity of researchers, subjects, and the research process itself is well protected. We have also built up a body of experience that will allow us to create good practice guidelines beneficial to all parties involved. Some of us, however, especially perhaps those new to the field, have not paid enough attention to the issues raised in this chapter. Yet, the inherent tensions between the fundamental methods of academic research and the dictates of private industry do not allow the luxury of neglect.

Note: Originally published in *New Solutions, 11*(4), pp. 295-306, 2011.

CHAPTER 14

Factors Influencing Ergonomic Intervention in Construction: Trunkman Case Study*

Scott Fulmer, Lenore S. Azaroff, and Susan Moir

This case study examines factors affecting the use of equipment designed to prevent lower back strain in laborers who pour concrete on major highway construction sites. Qualitative methods of organizational analysis were used to characterize factors identified from interviews and participant observation.

The major obstacles to the use of the control on site were

1. Managers placing a low priority on ergonomics
2. Safety officers' limited power in organizational hierarchies
3. Rationalizing, rather than challenging, resistance to change
4. Lack of a forum to share knowledge about interventions

Several organizational factors impeded the adoption of a technically effective, low-cost safety control on the site studied. The implementation of the control ultimately resulted from actions taken by the investigators, suggesting that safety programs present at the site are not always adequate to realize feasible interventions.

Construction consistently reports among the highest injury and illness rates of any industry [1]. In response, the National Institute for Occupational Safety and Health (NIOSH) has funded the development of a wide range of construction

*Contract Grant Sponsor: The National Institute for Occupational Safety and Health through The Center to Protect Workers' Rights. Contract grant number: CCU317202.

safety and health interventions [2]. While information about these interventions has been widely disseminated [3], the industry has been slow to adopt them.

This fact is consistent with observations that the construction industry resists the intrusions of outsiders and opposes change even while it demands better interventions [4]. Many construction companies lack resources and incentives to improve health and safety. Construction managers are responsible for finishing jobs on time and within budget and have little time to think about innovation, while costs associated with employees' health problems are frequently shifted to health insurance, Social Security, or Medicaid [5].

A key challenge to construction safety and health research is therefore the translation of controls that efficaciously reduce hazards under controlled conditions into interventions that are effective, i.e., that reduce hazards on real work sites. This process is influenced by the social organization of the construction workplace, about which little is published.

This study investigated the adoption of a control used to prevent low back injuries associated with the trunkman task, which consists of guiding the hose used to pour liquid concrete into forms. Ergonomic job analyses [6] reinforced other studies' findings [7] in identifying the trunkman task as a source of high exposure to risk factors for sprains and strains, including unstable footing, frequent heavy lifting, and static loading to the lower back and upper extremities. Participants in the Construction Occupational Health Program's (COHP's) worker training program also had identified this task as a priority for ergonomic improvements.

This case study describes an attempt to introduce a technical intervention called a "skid plate." Previous research had found that use of these plates secured to concrete-filled hoses significantly reduced the risk of low-back disorders by reducing low-back moments, lateral velocity, and sagittal flexion involved in manipulating horizontal hoses for concrete pours [7]. This research also found that the small sample of workers studied liked the control and thought that it decreased the exertion of pulling hoses.

In the present study, researchers from the COHP applied qualitative techniques of organizational analysis to evaluate the intervention and characterize the factors that limited or supported its effectiveness at the work site. Examples from other sites and intervention attempts are included as illustrative support.

MATERIALS AND METHODS

This study took place on a major highway construction site with more than 300 workers from eight trades employed by 26 subcontractors. Researchers and site personnel had developed relationships through years of field research. Limited data on the same task also were collected on a second, demographically similar site.

The study design was a case study of an attempt to introduce the use of skid plates. Classification of the factors influencing the effectiveness of the skid plate intervention was based on the previous five years of COHP fieldwork [8-11]. Extensive field notes were manually examined for factors identified by investigators, workers, or managers as affecting the adoption or maintenance of an intervention on a construction site or other aspects of its effectiveness at reducing hazards. More than 40 factors were identified [5].

For this case study, investigators produced extensive field notes on the trunkman task, associated hazards, and all activities related to introducing the intervention to the site. They specifically sought information related to factors previously identified. These notes were entered into a QSR NVivo [12] text database and coded according to the factors and other categories that emerged.

Field notes were coded in parallel by the principal field researcher, a second COHP researcher with extensive field experience on the research site, and the principal coder. Discrepancies were identified and codes further clarified or modified to a consistent, reproducible system.

When themes were identified in this case study, they were triangulated with data from previous COHP activities. These previous data were used to reinforce or challenge researchers' perceptions of issues important to the introduction of interventions in the context under study.

RESULTS

The Trunkman Task

Workers lift and drag the hose ("slickline" or "snake") with "canes" made of reinforcing bar. The first worker performs the most strenuous work, often supporting the end of the hose in a static posture. Measurements with a force gauge and wire demonstrated that lifting any part of a full hose pinned the force gauge at its maximum value of 112 pounds. Researchers calculated that the entire 10′ 5" diameter hose segment, including a connecting clamp, weighed just less than 300 lbs [6].

The hose sometimes had to be lifted over a form insert extending about twelve inches over the rebar mat. Researchers observed that three men who used the hooks to lift the hose over the insert became visibly exhausted, needing several moments to recover. Once they observed one man lift the hose over the same insert by himself.

Despite the forceful exertion required to perform the trunkman task, a member of a concrete pour team and his foreman told researchers that workers who complain about the difficulty of this job "don't know what they're doing." More than once they stated, "If you want to know how to handle the snake, you have to think like a snake," explaining that allowing the slickline to wind like a snake reduces forces on the back. That is, to move the hose end in a given direction workers can move the far end in the opposite direction, a counter-intuitive approach.

Identification and Acquisition of the Control

In February 2001, a laborer on the concrete pour crew described to a researcher and drew a tool that he called a "pan" that helps the snake slide more easily across the rebar mat. He reported that a friend had bought three and liked them, and named two vendors that sell them. The foreman on a second COHP study site had witnessed these pans, or "skid plates," in use on a neighboring contract, was impressed, and had his supervisor order some. He had at least 10 onsite within a week. After trying them, the foreman reported that he would not start a pour without the pans being set up or at least nearby and handy for use.

Researchers attending site safety committee meetings at the primary study site had repeatedly proposed ergonomic interventions and now announced their intention to perform an ergonomics evaluation of this tool. The site Safety Director obtained an unexpected increase to his department's budget and in March 2001 informed researchers that he would buy skid plates to evaluate for potential ergonomic benefits. He instructed a site safety officer to purchase three. As of May, the pans were not purchased. The safety officers repeatedly expressed their desire to obtain the pans, and one stated that he had the authority to make purchasing decisions for ergonomic tools. Nevertheless, they reported that no progress had been made in the acquisition process. Later in May a safety officer stated that the equipment had been ordered and would be in the field the following Friday. The pans had not arrived by June.

Three months after the safety director's original commitment to purchase the pans, a researcher took the initiative to speak to the site's purchasing agent, who called a vendor to order the equipment and request free delivery. The vendor agreed to supply the pans and delivery for free.

Use of the Control

When the pans arrived, a supervisor raised concerns that they might scratch the epoxy coating on the rebar mat. He added that the slickline used in the current pour was too short to use the pans. A foreman again explained to the researcher methods for handling the slickline, apparently assuming that the researcher did not understand the purpose of the pans.

When the researcher reminded the supervisors that the pans might ease one of the most demanding tasks on site, the foreman and supervisor began to discuss them. The foreman suggested that it would be helpful if the pans had wheels and higher cradles. The supervisor agreed to try the pan, noting that it could easily be removed if in the way. The foreman suggested that three pans might help the trunkmen and would be worth trying.

Researchers observed some incorrect use of the pans. Following an explosion of a pour hose plugged with dried concrete, the contractor had introduced a policy to tie hoses to the mat to prevent violent recoil until the concrete flowed freely. A foreman inadvertently tried to move the hose with the skid plate while the hose itself was still tied to the mat.

Researchers also observed a crew use a skid plate as dunnage to prevent a hose from moving and damaging the epoxy coat of the rebar. A supervisor expressed satisfaction with this use of the plates, which contradicted their intended purpose of helping the hoses to move more easily. In addition, the hose was not secured when the skid plates were untied to move the hose. The skid plate allowed the hose to move more easily, accelerating rather than dampening the reverberation. This became a safety hazard.

A laborer who was not a regular on the pour crew set them up without tying the hose into the plate. He expressed dissatisfaction with the plates' effect. This response was consistent with findings by Hess et al. [7] that, although the plates' manufacturers do not explicitly recommend securing the plates to the hose, they are effective only when secured. The researcher instructed the foreman, who in turn instructed the laborer, to tie the hose in with wire.

Shortly afterwards a laborer successfully used the second skid plate to pull two lengths of the hose by himself. Another laborer commented, "They work pretty good. They need another one."

In April 2003 researchers contacted the site supervisor to determine whether the pans had continued in use on that site. He explained they had used them three times and attempted to tie the slickline onto the skidplates in three different ways that were ineffective. He added that the supplier did not supply them free of charge as the purchaser had implied they would, but that the general contractor had to pay for them.

Obstacles to Use of the Control

The following four key themes arose from the qualitative data coding process.

Low Priority of Ergonomics Interventions

Although the safety staff agreed that the pans were desirable, no immediate incentives motivated their purchase. During the same period the staff were addressing other threats to physical safety and compliance demands for the reduction of silica dust levels. Use of skid plates, on the other hand, was neither required by law, enforced by government agencies, nor necessary to the prevention of death or immediate life-threatening injuries.

Researchers observed that employers on the study site typically responded quickly to immediately life-threatening hazards. Most workers interviewed about this issue stated that they felt free to report imminent dangers to a steward, foreman, or managers. A steward stated that, unlike other employers he had experienced, "this contractor has *never* been unwilling" to correct serious hazards that he indicated to managers.

On the other hand, several previous observations had revealed that the employer sometimes avoided intervening in health and safety issues. Researchers observed an excavator listing in an apparently hazardous manner in full view of managers who took no action. Researcher observations and steward statements revealed that the contractors neglected basic housekeeping, resulting in frequent trips, slips, and falls. Discussions between the stewards and contractors failed to improve this situation.

Some managers and employees explained that this apparent neglect of certain hazards resulted from lack of financial incentives to ameliorate them. The project owner directs the contractor joint venture to adopt certain safety practices, but can only delay, not reduce, payment if the employer fails to comply.

A safety officer added that "Wrap-around makes safety a complete reactive job." "Wrap-around" refers to the workers' compensation insurance policy on this job: the contractors insure themselves via their joint ventures, which cease to exist at the end of the project. Safety problems therefore do not increase insurance costs for the contractors in this or subsequent projects.

Safety Officers' Position in Site Hierarchies

One safety officer expressed commitment to the health and welfare of the workers but frustration with answering to two supervisors. He was accountable to both the project manager and the corporate safety director, either of whom could fire him. He was required to collaborate with the project manager in setting safety goals for the site, but felt he received little support in reaching those goals. Unlike his supervisors, he was not employed by the lead partner in the joint venture, but had been hired for a few months by a minor partner. Thus, not only were his supervisors politically split over production versus safety, but also his corporate allegiance was primarily to a company that had secondary power within the management team.

A safety engineer told researchers that building relationships on site takes a long time and is a prerequisite to instituting any change. The absence of such productive relationships prevented change in this case. Safety officers and a high-ranking production supervisor repeatedly asserted that the officers feared the supervisor, he had little respect for them, and they were unable to maintain a productive working relationship.

When safety officers perceived that this supervisor and perhaps other managers were opposed to using the skid plates, they seemed afraid to pursue the purchase. One implied that the supervisor had interfered with efforts to purchase the pans and that he and the supervisor had entered into a running conflict over the issue.

In practice, therefore, purchasing and introducing the skid plates required approval from both the safety director and the site hierarchy of foremen, managers, and site supervisor. Any individual in this chain could expedite the purchase of the skid plates by taking action (observed on the white board in the safety office

"I'd rather be fired because of my actions rather than my inactions, so get the fuck out of my way"); conversely, any individual could prevent the process by failing to pass the request up to the next level for approval. After approval was granted at the highest levels, actual purchase and introduction of the control required actions by some of these same individuals. Anyone could stop this process by resisting or simply waiting until the concrete pours were completed.

Tension between safety officers and supervisors emerged in a variety of ways. A steward described one safety manager: "[He] is a good safety manager . . . . [He] told [a higher level manager] 'Fuck you'" because the manager was preventing him from doing his job well." A manager involved in site safety told a researcher about his frustration at his job. He felt under attack from all sides, including the owners, the unions, and OSHA, though he is not to blame for the problems on site. Additional frustration resulted from turnover: he reported that only one manager besides himself had been involved in the site's safety since the start of the job. When participants at a joint labor-management-researcher meeting raised the safety officers' inability to address a particular hazard, one officer stated, "I take it very damn personally." He was acknowledging the fact that the situation made him appear unable to perform his job. He explained, "We get negative support from management."

The apparent inability of safety officers to take necessary actions sometimes led to workers' distrust. In one instance of an acute air quality problem, workers refused to contact safety officers themselves. They demanded that their steward bring an officer because they did not trust the safety staff to take proper action in the absence of someone with authority to protect them.

Rationalizing Resistance to Change

Foremen and supervisors resisted experimenting with the skid plates by repeatedly asserting that they would not work, in some cases before hearing an explanation of their use. A negative response of one seemed to encourage the others to dismiss their use as well. One foreman explained, "Some experienced guys would rather not use the hose tool. You can try to help somehow and you end up making it worse." Some expressed concern that the pans might damage the epoxy or waterproofing on the rebar.

Such concerns appeared spurious. Skid plates in use on a secondary study site did not compromise the membrane for waterproofing. On the main study site, a supervisor and foreman agreed that the pans would damage the epoxy no more than the hose clamps and concrete. A field engineer on the study site decided that the pans would not hurt the waterproofing since he was unable to scratch it with his pocketknife.

Researchers perceived that managers were seeking reasons to avoid changes in plans and routines that might risk potential damage to the pace or quality of the work. In addition, one skeptical supervisor often appeared defensive and reacted against people whom he saw as outsiders questioning site procedures and thereby potentially threatening his authority. This supervisor was a young man who had risen to his position through formal education rather than experience in the building trades, and therefore experienced lack of respect from many workers.

In another example, a researcher observed a member of the concrete pour team discuss a method for reducing the frequency of twisted ankles on the rebar mat by using more rebar to make a smaller mesh. A high level manager present at the discussion immediately proposed a reason that this wouldn't work, and offered an alternative method. The manager's suggestions were based on his knowledge of engineering specifications and experience on other sites. The researcher perceived that the manager opposed the worker's suggestion in order to preserve his authority. In any case, the manager's suggestion, though practical, served to stop the worker's input into the conversation. Ultimately, neither suggestion was implemented.

Lack of a Forum to Share Knowledge about Interventions

Although some members of the pour crew were familiar with skid plates used on another site, and were even able to identify vendors, no existing mechanism allowed them to transmit this information to people with the authority to act on it.

Another example of this phenomenon arose for another control used to ease the trunkman task. Pour crews on a secondary study site used an articulated, counter-weighted boom to place concrete. Ergonomic job analyses [6] demonstrated that this tool eliminated the trunkman's extremely forceful exertions to lift and drag the slickline, but may have introduced static loading.

A laborer on a pour crew had heard of the boom, he thought possibly in *Laborer* magazine, and thought it would be "beautiful." The laborers' steward was unfamiliar with this tool, and no managers were observed to promote its acquisition on the principal study site, where to researchers' knowledge none was ever used.

In an effort to establish a forum for exchanging this type of knowledge, investigators attempted unsuccessfully to obtain management permission to establish a standing labor-management ergonomics committee. To date, no mechanism exists to consistently permit the introduction, evaluation, and maintenance of ergonomics interventions.

RELIANCE ON RESEARCHERS TO FACILITATE INTERVENTIONS

The lack of mechanisms to disseminate knowledge and promote interventions contributed to a reliance on COHP researchers to facilitate change. As part of the effort to obtain skid plates, a safety officer urged researchers to speak to a supervisor about the upcoming concrete pour schedule and use that opportunity to ask why the pans had not been purchased. The safety officer was himself unwilling to raise this issue with the supervisor but implied that the supervisor was responsible for the delay in the purchases. The officer also maintained that he was unable to obtain the pans without collaboration by the purchasing agent and brought a researcher to discuss the issue with the agent.

The skid plate project was one of several in which individuals on site had requested researcher assistance in shortcutting the chain of command for introducing interventions. A field engineer had asked a researcher to approach a safety director with a request to purchase anti-vibrational gloves to protect workers from the intense vibrations created by power tools. The safety director responded positively to this suggestion. In another case, a safety officer told the researcher that he had to act on safety interventions "to get him [the researcher] off his back."

DISCUSSION

Consistent with earlier findings [7], skid plates proved to be readily available, inexpensive, feasible tools that can reduce exposure to sprain and strain hazards during concrete pours. According to one manufacturer, they sell "in streaks." Their failure to take root in the context studied illustrated important issues that can affect occupational health interventions in the construction industry.

Some of the obstacles encountered were technical and particular to the skid plate control. Workers on two sites, Oregon [7] and Massachusetts (this study), had difficulty with the coupling of the skid plate to the hose. A manufacturer of skid plates said that they are sold without couplings or attachments because earlier models with attachments had caused problems for the user and did not sell well. He also felt that the major advantage of the skid plate, rather than reducing friction, is the prevention of the hose coupling from catching in the rebar mat.

Other obstacles to introducing the skid plates grew from the work organization on site. Key organizational factors included managers placing a low priority on ergonomics; safety officers' limited power in organizational hierarchies; rationalizing, rather than challenging, resistance to change; and lack of a forum to share knowledge about interventions. These findings were consistent with previous, unpublished, COHP experiences.

Other COHP field observations reinforced the idea that contractors and their safety staff place a low priority on ergonomic interventions. For example, researchers had earlier attempted to introduce an impact wrench to decrease forceful, awkward, and repetitive movements for ironworkers who insert reinforcing bars into Dowel By Substitution (DBS) holes embedded in a slurry wall. An ironworker foreman reported that it worked well, but when technical difficulties developed, a contract engineer decided that it was not important to try modifications recommended by the foreman to salvage its use.

Safety officers by themselves typically lack the power to perform needed interventions. The introduction of controls often requires both permission and action up and down chains of command. In some cases, implementation also requires each of these individuals to initiate purchase, acquisition, and introduction of a control even after an intervention is approved. The awkward hierarchical position of the safety officer was observed in the ergonomic intervention attempted for the DBS job. When the impact wrench required repairs, a field engineer told researchers that he could have obtained the special parts needed by using his connections to skip some links in the chain of command, but ultimately had rejected that approach to avoid creating resentment. When asked about using a safety officer for support, he stated unequivocally that the officer's involvement would hamper, rather than help, the process. Others on site also expressed frustration about the safety officers' perceived powerlessness. Stewards discussed several small safety problems and the current responsible site safety person: ". . . and [he's] sitting on his ass! I told him about [the safety problem], and he said, 'I can't do anything about it.' And I said, 'Well, you're the fucking safety guy!'"

Once the chain of command has been activated for approving an intervention, each person on the chain may perceive an incentive to avoid risk by not taking the next step. They may offer rationalizations for avoiding change, which particularly in construction can result in delaying an intervention until a job is completed, or until an individual or crew becomes injured or ill.

Previous COHP observations were consistent with this study's finding that managers or foremen may delay interventions or quickly identify reasons to reject them. At times, they may redirect discussions of hazards to fixing the blame for their existence rather than implementing solutions. These responses appear to avoid the risks inherent in changing ongoing processes in ways that could potentially affect costs, efficiency, or productivity.

For example, a researcher discussing the site's repeated incidents of overexertion injuries associated with lifting walers told him about the "Binford Crab" tool for easing this task [13]. The manager immediately gave reasons that this tool, invented and used by piledrivers, wouldn't work. A steward responded with an explanation of why these concerns were misplaced. Relationships between safety officers and supervisors are critical to starting the intervention process. Researchers in this case observed a deadlock between safety staff and other managers that helped prevent the implementation of even simple, low-cost controls if they exceed compliance requirements. Tensions between safety officers and other managers have frequently appeared to prevent successful completion of this process in other cases.

Over ten years of fieldwork, COHP researchers have identified no mechanism to reproduce interventions that have been effective at other work sites or used by other employers. Interventions identified toward the bottom of the hierarchy, for example at the safety officer level, must endure this entire process from the beginning. Likewise, there appears to be no mechanism for workers to communicate their knowledge of controls to the decision-makers. Unionized workers can inform their stewards in case of imminent danger, but are not encouraged to make preventive suggestions.

The decision-makers, high-level supervisors, often lack the practical working experience to recognize hazards, especially sprain and strain hazards, requiring intervention. This issue appears to be exacerbated by recent trends toward hiring managers based on college education rather than trades experience.

Throughout the COHP program, stewards observed that the research provided an important, otherwise absent, forum for management and labor from various trades to listen to each others' concerns. For example, the "Ironworkers' Box" (14) is a simple control consisting of a wooden box that ironworkers can stand on to prevent working overhead to insert rebar. Constructing these boxes is, technically speaking, a trivial task for a skilled carpenter. However, workers in other trades typically do not directly request jobs of carpenters (at times foremen do arrange small favors between trades). Implementation of this intervention would require initiation from a steward or foreman who would obtain permission from the supervisor, who would then proceed to the site superintendent. If the superintendent granted permission, he would direct a site engineer to implement the intervention. The engineer would contact a carpenters' foreman, who would instruct the carpenters to build the box. Ironworkers themselves would then be responsible for using and maintaining the control.

In the case of the Ironworkers' Box, this complex series of steps was abbreviated by the presence of researchers who brought together members of different trades employed by different subcontractors and facilitated their communication with upper level management. It seemed improbable that the control would have been attempted otherwise.

These earlier observations suggest that, in this study, a more consistent researcher presence on site might have led to effective use of the skid plates. Daily inquiries about purchasing the plates may have motivated action, while researcher support may have tipped the balance between safety staff and managers in the quiet struggle over implementing this intervention. Organizational structures intrinsic to the construction site studied were not, however, adequate for implementing a relatively straightforward and potentially health- and money-saving intervention.

This case study investigated just one highway construction site and drew on additional findings from other sites similar in time, geography, and type of work. The degree to which its findings may be generalized to other construction worksites is not known. Nevertheless, these findings indicate that further work is crucial to identify and develop structures or systems that facilitate, rather than block, measures to prevent injuries to construction workers.

ACKNOWLEDGMENTS

The authors wish to acknowledge the many people who committed time, thought, and hard work to this study. In particular, the stewards of the New England Regional Council of Carpenters Local 33, Laborers International Union of North America Local 223, International Union of Operating Engineers Local 4, and the International Association of Bridge, Structural, Ornamental and Reinforcing Ironwork Local 7, were involved daily in the effort to protect and promote health and safety among their brothers and sisters on site, and their cooperation and presence assured our effectiveness. Several construction companies—Bechtel/Parsons Brinckerhoff, Slattery Skanska, Inc., JF White Contracting Company, Perini, Rusco Steel, Regis Steel—generously opened sites to observation or made significant contributions through their resources to allow critical examination of the struggle to promote health and safety. Many graduate students of occupational health at the Work Environment Department at the University of Massachusetts Lowell also made invaluable contributions to this study. The Center to Protect Workers' Rights provided funding and critical interest in this chapter.

REFERENCES

1. The Center to Protect Workers' Rights. *The Construction Chart Book: The United States Construction Industry and Its Workers.* 3rd Ed. Silver Spring, Maryland: The Center to Protect Workers' Rights, 2002.
2. The National Institute for Occupational Safety and Health. *A Compendium of NIOSH Construction Research 2002.* 2003. Washington, D.C.: National Institute for Occupational Safety and Health Publication No. 2003-103, 2003.
3. S. Schneider, L. Punnett, and T. Cook. Ergonomics: Applying what we know. In: K. Ringen, A. Englund, L. Welch et al., eds. *Occupational Medicine: State of the Art Reviews*, 10, 2, Construction Safety and Health, 385-394, 1995.
4. K. Ringen and E. J. Stafford. Intervention Research in Occupational Safety and Health: Examples from Construction. *American Journal of Industrial Medicine, 29*:4, 314-320, 1996.
5. S. Moir. *Worker Participation in Occupational Health and Safety Change in the Construction Workplace.* Doctoral dissertation: Work Environment Department, University of Massachusetts Lowell, 2004.
6. S. Fulmer. *Ergonomic Intervention in Concrete Pouring.* Unpublished paper: Work Environment Department, University of Massachusetts Lowell, 2004.
7. J. A. Hess, S. Hecker, M. Weinstein, and M. Lunger. Ergonomics in Construction: An Intervention with Concrete Laborers. International Society of Biomechanics XIXth Conference Proceedings, Dunedin, New Zealand, 2003.
8. S. Moir and B. Buchholz. Emerging Participatory Approaches to Ergonomic Interventions in the Construction Industry. *American Journal of Industrial Medicine, 29*:4, 425-430, 1996.
9. B. Buchholz and S. Moir, HealthTrak: A Participatory Model for Intervention on Ergonomic and Other Health Hazards in Construction. Final Report on Grant Number

R01 OH113060 from the National Institute for Occupational Safety and Health of the Centers for Disease Control and Prevention. Cincinnati, Ohio, 2002.

10. S. Moir, B. Buchholz, and J. Garrett. HealthTrak: A Participatory Model for Intervention in Construction. The International Ergonomics Association 13th Triennial Congress. Tampere, Finland, 1997.
11. B. Buchholz, S. Moir, and J. Garrett. HealthTrak: A Participatory Approach for Ergonomic Improvement in Construction, Second International Conference of CIB Working Commission W99, Honolulu, Hawaii, 1999.
12. NVivo qualitative data analysis program, version 2. QSR International Pty Ltd. Melbourne, Australia, 2002.
13. The Construction Occupational Health Program. The Binford Crab Clamp. Bright Ideas #4. University of Massachusetts Lowell. Lowell, Massachusetts, 1999.
14. The Construction Occupational Health Program. Ironworkers Box. Bright Ideas #7. University of Massachusetts Lowell. Lowell, Massachusetts, 2002.

Note: Originally published in *New Solutions, 16*(3), pp. 235-247, 2006.

CHAPTER 15

Green Chemistry in California: A Framework for Leadership in Chemicals Policy and Innovation

Michael P. Wilson, Daniel A. Chia, and Bryan C. Ehlers

Special Report to the California Senate Environmental Quality Committee and Assembly Committee on Environmental Safety and Toxic Materials, March 2006, From the California Policy Research Center and the Center for Occupational and Environmental Health, University of California.

This chapter summarizes a University of California chemicals policy report commissioned by the California Legislature. The report makes the case that long-standing weaknesses in the Toxic Substances Control Act have produced a flawed chemicals market in the U.S. that "undervalues" the hazardous properties of chemicals relative to their function, price, and performance. These market conditions have dampened industry interest in cleaner chemical technologies, such as green chemistry. A new U.S. chemicals policy will need to improve the flow of chemical information; enhance the capacity of government to control chemical hazards; and increase public investments in green chemistry research and education.

The UC report formed the basis for two new laws in California, AB 1879 (Feuer, D-LA) and SB 509 (Simitian, D-Palo Alto), and formed the basis for the state's Green Chemistry Initiative, a first-time effort by California EPA to craft a comprehensive approach to identifying, prioritizing, and taking action on chemicals of concern. The full report is available at http://coeh.berkeley.edu/docs/news/06_wilson_policy.pdf. The report's analysis and key findings appeared in August 2009 in *Environmental Health Perspectives* (Wilson and Schwarzman, Toward a New U.S. Chemicals Policy: Rebuilding the Foundation to Advance New Science, Green Chemistry and Environmental Health, Vol. 117(8), 1202-1209)."

By 2050, California's population is expected to grow by about 50%, from 36 to 55 million residents. This expansion will be accompanied by a growing set of social, economic, and environmental problems whose *magnitude* will be determined in large part by the policy decisions California makes now and in coming years. In charting a course to a sustainable future, policymakers will need to guide industrial development in such a way that it fully integrates matters of environmental quality and human health. In practice, if California is to create a future characterized by improving social, environmental, and economic conditions, industrial development will need to solve, not exacerbate, the public and environmental health problems facing the state today. To move California in this direction, policymakers need the support of research that links the science of public and environmental health to innovative policy solutions. The report summarized here serves that purpose in the area of chemicals policy.

This report makes the case that a modern, comprehensive chemicals policy is essential to placing California on the path to a sustainable future. Problems associated with chemicals are already affecting public and environmental health, business, industry, and government in California. On the current trajectory, the coming years will see these problems broaden and deepen. Correcting these problems will require much more than isolated chemical bans and other piecemeal approaches that currently characterize the Legislature's efforts in this arena. Rather, a comprehensive approach is needed that corrects long-standing federal chemicals policy weaknesses and builds the foundation for new productive capacity in *green chemistry*—the design, manufacture, and use of chemicals that are safer for biological and ecological systems. This approach to chemicals policy will link economic development in California with improved health and environmental quality, but it will require a long-term commitment to leadership on the part of California policymakers.

We describe initiatives by leading California businesses and the European Union (E.U.) that are already driving interest by industry in cleaner technologies, including green chemistry. Given California's unparalleled capacity for innovation and its scientific, technical, and financial resources, a proactive response to these developments in the form of a modern, comprehensive chemicals policy could position California to become a global leader in green chemistry innovation. The report illustrates that to do so, California will need to adopt a chemicals policy that greatly improves chemical information, regulatory oversight, and support for green chemistry research, development, technical assistance, and education.

METHODS

We used four research methods in preparing this report: a literature review, interviews with key informants, participation in chemicals policy meetings, and peer review. Over a two-year period, the primary author held discussions with chemicals policy experts affiliated with academic institutions, scientific bodies, governmental agencies, chemical producers, downstream users of chemicals, entities within the European Union, small- and medium-sized enterprises, environmental organizations, and labor organizations. In addition, between April 2003 and February 2006, the primary author participated in 35 meetings and conferences pertaining expressly or in part to chemicals policy matters; he presented the report's key concepts at 17 of these meetings. The report reflects feedback produced throughout this process.

MAJOR FINDINGS

The scale of chemical production is immense and will continue to expand globally.

Every day, the U.S. produces or imports 42 billion pounds of chemicals, 90% of which are created using oil, a non-renewable feedstock. Converted to gallons of water, this volume is the equivalent of 623,000 gasoline tanker trucks (each carrying 8,000 gallons), which would reach from San Francisco to Washington, D.C. and back if placed end-to-end. In the course of a year, this line would circle the earth 86 times at the equator. These chemicals are put to use in numerous processes and products, and at some point in their life cycle many of them come in contact with people—in the workplace, in homes, and through air, water, food, and waste streams. Eventually, in one form or another, nearly all of them enter the earth's finite ecosystems.

Global chemical production is expected to double every 25 years for the foreseeable future. Between now and 2033, the U.S. EPA expects 600 new hazardous waste sites to appear each month in the U.S. and require cleanup, adding to 77,000 current sites. Efforts at site mitigation are expected to cost about $250 billion. Given the scale, pace, and burden of chemical production, the toxicity and ecotoxicity of chemicals are of great public importance.

Many chemicals that are useful to society are also hazardous to human biology and ecological processes.

There is growing scientific concern over the biological implications of chemical exposures that occur over the human lifespan, particularly during the biologically sensitive period of fetal and child development. Hundreds of chemicals that are released into the environment are accumulating in human tissues; the U.S.

EPA found just under 700 such chemicals in a nationwide survey of Americans in 1987. Many of these chemicals enter the developing organ systems of fetuses and infants through the maternal bloodstream and through breast milk. Animal studies indicate that some can interact with and disrupt the development of these systems, such as the endocrine system, at very low doses. Among children, chemical exposures are estimated to contribute to 100% of lead poisoning cases, 10% to 35% of asthma cases, 2% to 10% of certain cancers, and 5% to 20% of neurobehavioral disorders.

Occupational disease continues to exact a tremendous toll in California. Each month, an estimated 1,900 Californians are diagnosed with a preventable, deadly chronic disease that is attributable to chemical exposures in the workplace; another 540 Californians die as a result of a chronic disease linked to chemical exposures in the workplace. The U.S. Occupational Safety and Health Administration (OSHA) has adopted workplace exposure limits for only 193, or about 7%, of the 2,943 chemicals produced or imported in the U.S. at more than one million pounds per year. Immigrants, minorities, and lower-income groups—as workers and as residents—are at particular risk of exposure to hazardous chemicals.

There are extensive deficiencies in the federal regulation of chemicals.

Of all federal environmental statutes, the Toxic Substances Control Act of 1976 (TSCA) is the only law that is intended to enable regulation of chemicals both before and after they enter commerce. However, studies conducted by the National Academy of Sciences (1984), the U.S. General Accounting Office (1994), the Congressional Office of Technology Assessment (1995), Environmental Defense (1997), the U.S. EPA (1998), former EPA officials (2002), and the U.S. Government Accountability Office (2005) have all concluded that TSCA has not served as an effective vehicle for the public, industry, or government to *assess* the hazards of chemicals in commerce or *control* those of greatest concern.

- The TSCA inventory lists 81,600 chemicals, 8,282 of which are produced or imported at 10,000 pounds or more per year.
- TSCA does not require chemical producers to generate and disclose information on the health and environmental safety of these chemicals—or on the approximately 2,000 new chemicals that enter the market each year. The result is that there is an enormous lack of information on the toxicity and ecotoxicity of chemicals in commercial circulation.
- TSCA places legal and procedural burdens on the EPA that have constrained the agency's capacity to act. Since 1979, the EPA has used its formal rule-making authority to restrict only five chemicals or chemical classes, though the agency reported in 1994 that about 16,000 chemicals in the U.S. were of some concern on account of their structure and volume in commerce.

- TSCA has not provided a vehicle for channeling federal support to research in cleaner chemical technologies, including green chemistry.

Voluntary initiatives on the part of the chemical industry to correct some of these weaknesses are positive but do not make up for TSCA's structural weaknesses. Other federal laws that pertain to chemicals are essentially end-of-pipe statutes that do not allow for review of chemicals prior to their introduction into commerce. Together, five major federal statutes apply to only 1,134 chemicals and pollutants. The weaknesses of TSCA and other federal statutes have produced three fundamental problems in the U.S., which we refer to as the chemical Data Gap, Safety Gap, and Technology Gap.

TSCA's weaknesses are adversely affecting California.

The chemical Data Gap, Safety Gap, and Technology Gap have created a broad set of problems for public and environmental health, industry, business, and government for U.S. states, including California.

The Data Gap

Without comprehensive and standardized information on the toxicity and ecotoxicity for most chemicals, it is very difficult even for large firms to identify hazardous chemicals in their supply chains. Along with consumers, workers, and small-business owners, they do not have the right kinds of information to identify safer chemical products. The lack of chemical information weakens the deterrent function of the product liability and workers compensation systems.

The Safety Gap

Government agencies do not have the information they need to identify and prioritize chemical hazards systematically, nor the legal tools to mitigate known hazards efficiently.

The Technology Gap

The lack of both market and regulatory drivers has dampened motivation on the part of U.S. chemical producers and entrepreneurs to invest in new green chemistry technologies. There has been virtually no government investment in green chemistry research and development.

Meanwhile, evidence of public and environmental health problems related to chemicals continues to accumulate. Each year, the California Legislature faces numerous bills related to public concerns over chemicals; on the current trajectory, the number of such bills is likely to grow. Correcting the chemical Data, Safety, and Technology Gaps engendered by TSCA will require a modern, comprehensive approach to chemicals policy in California.

Developments in the European union and among leading California businesses are driving interest in cleaner technologies, including green chemistry.

Facing a similar set of problems, the E.U. is implementing sweeping new chemicals and materials policies that are driving global changes in ways that will favor cleaner technologies, including green chemistry.

- The E.U. *Restriction of Hazardous Substances in Electrical and Electronic Equipment* (RoHS) Directive will prohibit the use of lead, cadmium, mercury, certain flame-retardant chemicals, and other toxic materials in electronic and electrical equipment sold in the E.U.
- The *Waste Electrical and Electronic Equipment* (WEEE) Directive requires electronics producers to take back their products at the end of their useful life.
- The *Registration, Evaluation and Authorization of Chemicals* (REACH) initiative will require chemical producers to register most chemicals that are widely used and will place restrictions on the use of about 1,400 chemicals of very high concern.

It is becoming clear that cleaner technologies will play an increasingly important role in industrial activity globally—among both developed and developing nations. The E.U. government's policies to motivate investment in cleaner technologies, though difficult for some E.U. producers in the short term, are expected to lead to a long-term E.U. competitive advantage in this arena.

Lacking similar government leadership in the U.S., a number of large businesses have been working independently to implement strategies for identifying hazardous chemicals in their supply chains and removing those chemicals from their operations. California businesses at the forefront of this effort include Kaiser Permanente, Catholic Healthcare West, Intel, Hewlett-Packard, IBM, Bentley Prince Street, Apple, and others. These developments signal a growing demand among U.S. businesses for safer chemicals and better chemical information; these efforts, however, are constrained by the Data, Safety, and Technology Gaps. Effective leadership in chemicals policy to close these Gaps is now called for in the U.S.

California needs a modern, comprehensive chemicals policy to address pressing public and environmental health problems and to position itself as a global leader in green chemistry innovation.

These developments have opened an opportunity for California to position itself as a leader in green chemistry science and technology. To do so, California will need to correct the Data, Safety, and Technology Gaps, which have given rise to conditions in the U.S. chemicals market that favor existing chemicals and discourage investment by chemical producers in new green chemistry technologies. Large "sunk" investments by industry in existing chemical technologies will make it difficult to transition to an industrial system based on cleaner technology, including green chemistry; this transition, however, will have to be made if California is to respond proactively to developments in the E.U. and address a host of chemical problems affecting public and environmental health, business, industry, and government in the state.

We propose three chemicals policy goals that will move California in this direction:

Close the Data Gap—Ensure that chemical producers generate, distribute, and communicate information on chemical toxicity, ecotoxicity, uses, and other key data.

Close the Safety Gap—Strengthen government tools for identifying, prioritizing, and mitigating chemical hazards.

Close the Technology Gap—Support research, development, technical assistance, entrepreneurial activity, and education in green chemistry science and technology.

Because many policy mechanisms could be employed to reach these goals, we recommend that as a first step the Legislature establish a chemicals policy task force to explore various mechanisms and develop a legislative proposal for a comprehensive policy based on the findings of this report. We recommend that the task force be charged with developing the proposal for the 2007 legislative session.

REPORT BACKGROUND, DEVELOPMENTS, AND WEB ACCESS

The California Senate Committee on Environmental Quality and the Assembly Committee on Environmental Safety and Toxic Materials commissioned the University of California (UC) Chemicals Policy report in January 2004. It was completed and released to these Committees in March 2006 by the California Policy Research Center of the UC Office of the President. A 13-member UC Advisory Committee provided technical oversight for the report. California Senator Joseph Simitian (D-Palo Alto) convened public hearings on the report in June and October 2006. Michael P. Wilson presented the report's key findings before the U.S. Senate Committee on Environment and Public Works on August 2, 2006. The full report and related links can be accessed at http://coeh.berkeley.edu/news/06_wilson_policy.htm

The California Legislature established COEH in 1978 (AB 3414) to improve understanding of occupational and environmental health problems in California and work toward their resolution through research, teaching, and service (http://coeh.berkeley.edu/). The California Policy Research Center (CPRC), under the aegis of the UC Office of the President, applies the UC system's research expertise to analysis, development, and implementation of public policy in

California (http://www.ucop.edu/cprc/#). CPRC, COEH, and the UC Toxic Substances Research and Teaching Program provided funding for the report.

The members of the UC Advisory Committee are: John R. Balmes, MD, Professor, School of Medicine, UC San Francisco, and Professor, Environmental Health Sciences, UC Berkeley; Carl F. Cranor, PhD, Professor, Department of Philosophy, UC Riverside; S. Katharine Hammond, PhD, Professor, Environmental Health Sciences, School of Public Health, UC Berkeley; Bill E. Kastenberg, PhD, Professor, College of Engineering, UC Berkeley; Ann Keller, PhD, Assistant Professor, Health Policy and Management, School of Public Health, UC Berkeley; Amy D. Kyle, PhD, MPH, Research Scientist, Environmental Health Sciences, UC Berkeley; Geoff Lomax, DrPH, Research Director, Environmental Health Investigations Branch, California Department of Health Services; Timothy Malloy, JD, Professor, School of Law, UC Los Angeles; Thomas E. McKone, PhD, Senior Staff Scientist, Lawrence Berkeley National Laboratory, and Adjunct Professor, Environmental Health Sciences, School of Public Health, UC Berkeley; Dara O'Rourke, PhD, Professor, Environmental Science, Policy and Management, College of Natural Resources, UC Berkeley; Julia Quint, PhD, Chief, Hazard Evaluation System and Information Service (HESIS), California Department of Health Services; Christine Rosen, PhD, Professor, Haas School of Business, UC Berkeley; David J. Vogel, PhD, Professor, Haas School of Business, UC Berkeley, and Professor, Department of Political Science, UC Berkeley.

The UC Centers for Occupational and Environmental Health at Berkeley and UCLA published a second report in 2008, *Green Chemistry: Cornerstone to a Sustainable California,* which was commissioned by California EPA and carried the signatures of over 125 UC faculty from seven campuses (see http://coeh.berkeley.edu/greenchemistry/briefing/default.htm).

ACKNOWLEDGMENTS

The authors would like to acknowledge the members of the UC Advisory Committee and Holly Brown-Williams, Al Averbach, John Crapo, Andrés Jiménez, Robert Spear, John Balmes, Steve Shortell, and the individuals in California, the United States, and the European Union who freely offered their expertise in chemicals policy.

Note: Originally published in *New Solutions, 16*(4), pp. 365-372, 2005.

CHAPTER 16

The Sustainability Solutions Agenda

Dan Sarewitz, Dick Clapp, Cathy Crumbley, Polly Hoppin, Molly Jacobs, David Kriebel, and Joel Tickner

SUSTAINABILITY AND SCIENCE

The challenge of reconciling societal aspirations and environmental limits is captured by the term "sustainability." There are good reasons to believe that society is on an unsustainable path for the longer term. And, while no one knows exactly what a sustainable path might look like, we think we know good ways to start down that path—reducing fossil fuel dependencies, for example. We are told nearly every day that global climate change is the most urgent consequence of our current unsustainable path, and a complex web of causes and effects can already be seen in recent spikes in oil and food prices all around the planet. These ripples in the global economy illustrate how everything really is connected to everything else, how the long-term consequences of actions taken in the present are very difficult to predict, how the social, environmental, and economic costs and benefits of even well-intentioned actions may be unevenly distributed, and how the momentum of a society committed to continual economic growth via competitive markets can be extraordinarily difficult to redirect.

Recognition of the urgent need to identify effective solutions to planetary problems like global warming has led to increased attention to studying the complex systems, like climate, from which these crises seem to arise. There are many efforts underway to define and develop "sustainability science," "systems science," and other related approaches [1, 2]. We are also exploring these themes, and in this chapter, we explain our view of how best to link knowledge and action to the quest for greater sustainability, and contrast what we have learned in the course of our scientific work with other perspectives.

Science is widely recognized as a crucial tool for moving toward a more sustainable world. This recognition flows from the assumptions that:

1. additional scientific understanding about society-nature interactions is necessary to define a path toward sustainability;
2. greater understanding will guide the decisions allowing society to follow that path; and
3. this understanding will also motivate people to make the behavioral changes necessary to act more sustainably.

In this view, which we think is implicitly and widely held, knowledge comes first, then action. The "knowledge first" doctrine is a core assumption of modern society, where rational action is viewed as deriving from factually correct assessments of the causes of a problem.

One example of a knowledge-first approach to sustainability is risk assessment (and related methods like cost/benefit analysis): first characterize the costs and benefits, and then make choices accordingly. Risk assessment is widely viewed as a necessary input into rational decision making for sustainability issues such as environmental health, and is enshrined in many policies that regulate chemicals and materials. Risk assessment has been justly criticized for being overly narrow—focusing on a single technology or chemical, and reducing social questions like: "do we want this technology?" down to the narrow "is there strong evidence that this technology is too risky?" [3].

While risk assessment is reductionist, other knowledge-first approaches have a systems level focus. "Sustainability science" was born from a critique of conventional science as not up to the challenge of confronting complexity. In strong contrast to risk assessment and its cousins, sustainability science "seek[s] to address the essential complexity" of human-environment interactions, recognizing that "understanding the individual components of nature-society systems provides insufficient understanding about the behavior of the systems themselves." Sustainability science demands "close collaboration between scholars and practitioners," and it aims at "creating and applying knowledge in support of decision making for sustainable development" [1]. This approach has been central to the strategy followed thus far for climate change, where a comprehensive international research program is intended to both inform and motivate a global transition away from fossil fuels. And, while we support the call for studying systems rather than individual risks, both risk assessment and sustainability science remain well within the bounds of the "knowledge first" paradigm.

RESEARCH FOR SOLUTIONS

We are setting out a different way to find pathways to a more sustainable society. Acknowledging the complexity of the planet and of human societies, we will draw a distinction between our approach and all "knowledge first" views of science for sustainability. What we call the Sustainability Solutions Agenda (SSA) responds to the reality that humans live in a technological world. Humans and their organizations produce and use technologies continually to accomplish important tasks with high reliability. At the same time, technologies play a central role in the important threats to sustainability. SSA recognizes and responds to these dual realities by focusing on the uses of technology in the real world, in the present, to ask: what opportunities exist for steering the design, production, and use of technologies away from unsustainable practices toward more sustainable ones, without sacrificing the value of these technologies?

SSA is thus at once visionary and pragmatic. On the one hand, it aims at a world where technologies are less harmful to humans and to nature, but on the other hand it assumes that often this vision can be most rapidly achieved through incremental introduction of alternatives and solutions without waiting for comprehensive knowledge of the relevant nature-technology systems. SSA of course recognizes that action must be informed by evidence. But rather than assuming that detailed systems knowledge will be the key to action, SSA recognizes that the possibility of positive change will motivate further change, without waiting for a convergence of people's values and interests. SSA seeks to identify a path for incremental political evolution toward sustainability in a world where political power is often concentrated in organizations, institutions, and corporations that are structured to resist such change.

SSA has the goal of helping humanity live sustainably on the Earth. It is program of integrated research and practice whose purpose is not simply to understand the world better, but to inform and motivate social actions toward sustainability.

SSA is informed by a long-term vision of technological change, but it is focused on uncovering paths to sustainability by improving current technological practice, applying admittedly limited knowledge to identify and evaluate technological alternatives. This practical and immediate emphasis is tempered by a commitment to continuous improvement in the future—and the recognition that some solutions will turn out to be mistakes, or dead-ends. It is inherently collaborative: the users of unsustainable technologies are sources of expertise that are essential to identifying feasible paths of change. Its orientation toward near-term solutions creates positive feedbacks and a sense of progress, empowerment, and shared mission. SSA integrates research and analysis with action aimed at social and technological change; these activities are inseparable and simultaneous. In our view, more knowledge is not a prerequisite for action; knowledge and action advance together. SSA therefore creates major challenges for institutions devoted to advancing sustainability, since many such institutions—universities especially—are structured around knowledge-first approaches.

THE MONTREAL PROTOCOL: AN EXAMPLE

An historical example may help to contrast SSA with knowledge-first approaches. The Montreal Protocol on Substances that Deplete the Ozone Layer is often cited as a signal success in applying systems-level knowledge to a problem of sustainability. The standard portrayal of this international agreement to phase out the production of chlorofluorocarbons follows the knowledge-first model: researchers discovered that CFCs depleted the Earth's stratospheric ozone layer, and as knowledge became increasingly certain, action ramped up to deal with the problem, culminating in an international agreement to phase out production signed by most of the world's nations. Different tellings of this story emphasize different aspects, such as diplomacy [4], corporate incentives [5], and scientific assessment activities [6]. But they all largely neglect a critic aspect of "the system:" CFCs provided essential functions as refrigerants and solvents upon which many sectors of society depended. These functions could not have been sacrificed without serious social disruption and financial losses for important economic sectors. Early actions, such as the phase out of CFC-propelled aerosol cans, were easy because alternatives for non-essential functions already existed; roll-on deodorants could substitute for spray cans for example. Such changes occurred even when scientific uncertainty about CFC impacts was high, but they created a sense of possibility and momentum, consistent with the SSA approach. A complete production phase out become practically and politically possible only when alternative chemicals serving the same essential functions (like keeping food cold and semiconductors clean) began to come on line. While the standard story is one of knowledge compelling action, the SSA perspective shows that decisive action only became possible once alternatives were identified, and that there was a dynamic relationship between knowledge and action that was more complex than first one, then the other.

Why is the CFC story conventionally related as one of action enabled by new knowledge rather than one of problem-solving enabled by technological substitution? The idea that problems are solved first by generating necessary new knowledge, and then taking rational action, is powerfully held in modern culture and is an enduring legacy of the Enlightenment. As we will discuss below, "knowledge first" is also an important justification for academia, with its emphasis on "pure" knowledge acquisition and academic freedom.

PROBLEM SYSTEMS VERSUS SOLUTION SYSTEMS

Sustainability science, in the knowledge-first mode, seeks to holistically characterize a problem in terms of its causes and mechanisms as a basis for subsequent action—the system of interest is that which contains the problem. SSA, in contrast, seeks to identify the possible pathways to solutions within the system—*the system of interest is that which contains the solutions.* This is a key distinction. For example, the industrial solvent methylene chloride is known to be toxic, but the mechanisms of its toxicity are not well understood. A knowledge-first approach might focus on developing a better understanding of how methylene chloride behaves in the environment and in the body as prerequisites for developing interventions; SSA focuses on understanding why and where methylene chloride is used as a basis for identifying alternative practices of production and use, and approaches to implementing those practices.

A more familiar example is global warming. The problem space for global warming is the climate system; the solution space is the global energy system. The latter is nested in the former and of course cannot be ignored, but the energy system (including energy consumers) must be understood in order to identify feasible, practical steps to reduce and mitigate global warming. By far the greatest focus of research on global warming, consistent with the knowledge-first mode, has been aimed at better understanding of the coupled atmosphere-ocean system and its links to the biosphere and society. The idea is that this research will motivate the right political responses. Yet if one views the problem from the perspective of the solution system—the energy system—alternative technological pathways toward greater sustainability have always been available, and the problem becomes one of how to motivate exploration of these pathways. The result of emphasizing knowledge of the problem system over action in the solution system is that we now find ourselves with strong evidence that the planet is warming, but years behind where we might have been in developing and disseminating solutions, had we take more seriously the political opportunities created by technological alternatives.

FROM KNOWLEDGE TO ACTION

SSA obviously depends upon scientific evidence, typically created in the knowledge-first mode, about the environmental and health implications of various phenomena, chemicals, materials, and practices. Moreover, SSA is not an argument against efforts to better characterize those implications. Rather, SSA is a mode of integrated inquiry and practice that moves from a domain of imperfect knowledge and ever-present uncertainty that always characterize a problem, to a domain of potential action based on the search for and availability of a solution. The need to act in the face of uncertainty has different implications for researchers depending on assumptions about the links between evidence and action. The "knowledge first" approach often ends with the publication of results in a scientific journal or report, without specifying possible avenues of action. This caution appears justified both because absolute proof is not possible, and because knowledge acquisition is conceived as separate from action. Again turning to the climate change example, the scientific assessment activities of the Intergovernmental Panel on Climate Change are intentionally insulated from implementation activities, for example those occurring under the UN Framework Convention on Climate Change. In bringing science and action together, SSA assesses the weight of evidence needed to draw conclusions, and views the potential for action in terms not just of the strength of evidence of harm or benefit, but also in terms of the ability to identify and implement potential solutions.

Lead, a neurotoxin and probable carcinogen, is still used pervasively in the production of electronic equipment, which in turn is the fastest growing source of waste in industrialized countries. If the challenge is to reduce the use of lead in the electronics industry, this means simultaneously conducting research to identify substances and processes that can substitute for lead solder and other uses of lead in electronics; working directly with the electronics industry to encourage substitution; working with policy makers to create incentives for substitution; and encouraging feedbacks among these and related activities to allow for learning and accelerated change.

Or if the challenge is to eliminate PCE, a volatile and toxic chlorinated hydrocarbon used in the highly decentralized, low tech dry-cleaning industry, this means developing pilot programs to demonstrate that alternative wet-cleaning technologies perform as well or better than dry-cleaning; creating an atmosphere where learning and technology adoption can proliferate among the thousands of small, locally owned establishments; and providing resources to aid in the initial investment in new wet-cleaning equipment for first adopters.

Or, if the challenge is to deal with a cancer cluster in a community, this may mean working with community members to address the most likely cause of the problem (for example, to seek alternatives to toxic chemicals used in a nearby factory) directly and in the short term, rather than engaging in lengthy epidemiologic studies which in any case are unlikely to find a "smoking gun" because the available research methods are still quite weak.

These thumbnail examples illustrate how SSA deals with the most vexing aspects of the sustainability challenge: complexity, *uncertainty about risk*, and *political conflict.* In essence, SSA avoids the obstacles created by these factors by focusing on available solutions, incremental change, and preservation of functionality. It eases the tension between risk assessment and precautionary approaches by moving the discussion away from risk uncertainties and onto the potential benefits offered by technological alternatives. It avoids never-ending demands for more knowledge about complex system behavior by focusing on clear paths of positive change within the larger system. And it engenders a shared sense of progress by focusing on incremental and measurable improvement.

DEALING WITH COMPLEXITY

SSA deals with *complexity* by focusing on unsustainable practices in society. SSA does not seek to fully characterize such practices within complex social and environmental systems. Rather it seeks intervention points where an activity (say, producing shoes) is tightly linked with an unsustainable practice (say, using volatile and toxic organic glues) for which more sustainable alternatives (say, using water-based glues) are either available or potentially available.

SSA deals with *uncertainty about risk* by moving discussions away from risk characterization and toward alternatives assessment [7, 8]. Complete understanding is never possible in characterizing the risk of a particular process, chemical, material, or practice. The continual existence of uncertainty supports competing views about acceptable risks, and sustains conflict about how to respond to risks (or even about whether the risk is "real"). But because SSA allows people and organizations to transition toward greater sustainability without sacrificing essential technological functions, it does not threaten the interests that depend on those functions. The question is not (for example) "how toxic is this chemical or process?" or "what are the mechanisms by which this chemical affects human health" but "how can we change from using this plausibly toxic substance to using a plausibly more sustainable substance that allows us to do the same job?" Thus, SSA deals with conflict by changing and often lowering the stakes associated with social change. The focus moves from characterizing a problem for which a person or organization is responsible, to specifying a solution that offers potential benefits (some of which may have been previously unrecognized) and manageable costs.

SSA allows for values to evolve toward greater attention to sustainability as a result of the positive experience of solving a problem. This process contrasts markedly with knowledge-first approaches to sustainability, which view scientific information (typically focused on proving the causes and magnitude of future impacts) as sufficient to convince people to take the right actions, even if those actions are perceived as against their immediate interests.

In finding a path to sustainability, societies must make extremely difficult decisions. When we arrive at one of these decision points, we often find that facts are uncertain, values are in dispute, the stakes are high, and decisions are urgent [9]. Under these conditions, conventional science is often too narrowly focused, slow, and overly cautious. Mistakes like the "Type III error" (providing an accurate answer to the wrong question) are common [10]. Also, the central question of causality ("how much global warming is due to anthropogenic greenhouse gases?" "how many people die from urban air pollution?") is often a stumbling block for traditional science because of its insistence on the existence of a single truth which science can, with enough resources, identify [11, 12]. SSA views these questions differently, asking not "does X cause Y?" but instead

Plausibly More Sustainable? A challenge for SSA is to develop robust principles of "plausibly safer." For example, the search for alternatives to hazardous chemicals and materials needs criteria for evaluating risks based on incomplete knowledge. An alternative might be considered an improvement over current practice if:

- the current chemical is persistent or bioaccumulative, and the alternative is biocompatible, biodegradable, or renewable; and
- the current chemical shows strong evidence of harm, and the alternative shows evidence of less harm, or shows little or no evidence of harm.

Because knowledge is always provisional, in most cases it will not be possible to entirely eliminate uncertainties about the benefits of potential solutions relative to current practice. Thus, in SSA:

- The alternative should always subject to future surveillance; and
- The alternative should be amenable to flexible production or future substitution as part of a process of continuous improvement.

"Given the possibility that X causes Y, is there a way to move toward more sustainable practice by replacing X while still preserving some or most of its benefits?"

Both SSA and sustainability science are committed to multi-stakeholder collaboration as an essential component of effective action. For sustainability science, collaboration between scientists and stakeholders aims at the production of knowledge that is "socially robust," meaning both technically sound and socially acceptable [13]. SSA, in contrast, is not centrally focused on knowledge production as a stimulus for action, but on stimulating concrete steps toward sustainability. Collaboration between SSA researchers and technology users (ranging from large firms to communities to individual households) aims directly at improving practice through technological substitution and changing contexts for decision making.

"Knowledge first" sustainability science is a product of the aspirations, organization, and social structure of the American research university and its commitment to creating new knowledge. University programs focusing on sustainability are notably more interdisciplinary than most other academic fields, yet they are still populated by professors and students whose job is to generate new knowledge that can advance their careers, largely through publication in peer-reviewed journals and grants from government agencies. Rational action motivated by rigorously produced knowledge is the model for how science stimulates social change, and sustainability science is the product of this culturally and professionally embedded view.

Trichloroethylene Reduction in Massachusetts. Trichloroethylene (TCE) is a solvent that has been targeted by US EPA and numerous states for replacement because of toxicity, including potential carcinogenicity, even as its human health effects remain disputed. TCE is commonly found at Superfund sites, and is particularly problematic because it can persist in groundwater and migrate into drinking water supplies. A main use of TCE is for degreasing metal parts, an application that creates a high risk of exposure to workers, especially in small firms that may be below the regulatory radar. These cleaning tasks can be performed with alternative organic solvents or with water-based cleaners. The latter are preferred because of the likelihood that the water-based cleaners are considerably safer and healthier than either TCE or any of the less toxic synthetic solvents.

The Massachusetts Toxics Use Reduction Institute (TURI), has been working with metal manufacturers to help them shift from TCE to safer cleaning solution [14]. TURI determined that a critical impediment to firms adopting safer alternatives was their concern that productivity and product specifications might suffer if they changed their standard metal cleaning procedures. To address this concern, the Institute built the Surface Solutions Laboratory specifically to evaluate the effectiveness of alternatives to TCE. TURI surveyed firms in Massachusetts that were potential TCE users to develop a roster of target firms, and then offered its services in assessing alternative parts cleaners. Assessments included tests at the Surface Solutions Lab using actual parts that the firms themselves needed to have cleaned, followed up in some cases by pilot projects involving on-site testing of alternatives in the firm's own facility. TURI also helped to develop cost-benefit analyses for alternatives, and worked with State agencies and professional organizations to demonstrate TCE replacement possibilities at workshops and meetings in an effort to reach more firms.

Working in Massachusetts and Rhode Island, TURI's efforts have thus far led to a 67% decrease in TCE use among cooperating firms, from a total of more than 280,000 lbs/yr to 95,400 lbs/yr [14]. This effort has also led to the reduction in use of other volatile cleaning solvents. TCE replacements included non-chlorinated solvents with no known health risks, and water-based, ultrasonic cleaning processes.

SSA IN ACADEMIA

In contrast, SSA requires innovative institutional arrangements that do not fit easily into existing organizational models for universities, or anywhere else for that matter. The skills required to implement the SSA agenda include, but go far beyond, those necessary for conventional academic research, and demand not just scientific expertise in relevant areas but strong organizing, training, and community-building skills. And the measures of success lie not in new knowledge created, but in real-world solutions achieved. These criteria offer a stark challenge to universities, whose model of professional success starts with academic researcher as producer of new knowledge, yet whose missions increasingly include a focus on contributing more effectively to societal well-being.

WHAT'S NEXT?

We have outlined an agenda for pursuing sustainability that starts with the recognition that humans are technological beings, and that people and organizations can make more sustainable choices about technological alternatives without sacrificing the functionality upon which they depend. SSA can thus help to create a new politics of sustainability, one catalyzed by the positive experience of incremental, beneficial change. While SSA neither obviates the need for more scientific knowledge, nor for new types of regulatory regimes, it does not demand that people change their behavior based on new knowledge or incentives, rather it highlights the options for positive change based on existing knowledge, laws, and technological alternatives. In this way, the path toward greater sustainability is discovered simultaneously with the evolution of both knowledge and values.

SSA as described here has been developed and is currently practiced at the Lowell Center for Sustainable Production (http://sustainableproduction.org/), and its partner institution, the Massachusetts Toxic Use Reduction Institute (http://www.turi.org/). An important short-term step for advancing the SSA agenda is to further develop easily communicated and relatively unambiguous criteria for making decisions about technological substitutions. A core principle of SSA is that the path toward greater sustainability can be discovered by substituting "plausibly more sustainable" technologies and practices for existing, unsustainable technologies and practices. Formalizing such criteria would enable and encourage the wider adoption of SSA. We have presented some possible criteria of "plausibly more sustainable" as applied to toxic chemicals, and explained why such criteria are likely to be less scientifically and politically contentious than more traditional risk-assessment-based frameworks for regulating technologies. However, what we have offered is at best a first step based on our own experiences and knowledge. An important next step is to convene international groups of practitioners working in the areas of alternatives assessment and technological substitution in areas ranging from toxic materials to agro-chemicals to energy technologies to develop a robust and expansive set of criteria for "plausibly more sustainable." The continued political gridlock in the U.S. over regulation and management of new and existing chemicals, materials, and processes provides strong motivation for pursuing this alternative framework. The twenty-fifth anniversary of the National Research Council's 1983 report *Risk Assessment in the Federal Government: Managing the Process* (the Red Book) reminds us of the continued need for fresh approaches to managing our technological ambitions.

Over the longer term, progress in advancing SSA can be promoted along paths of education and of public policy. For example, the idea that at any given time, many technological paths are available for achieving a particular desired functionality [15] is not a central aspect of science and engineering education—but it should be. The gradual rise of sub-fields like green chemistry, and of new ways to think about functionality, like biomimicry, speak to the potential for more sustainable technological alternatives, but they remain rather marginalized relative to conventional disciplinary approaches. Creating a culture that encourages scientists and engineers to explore multiple paths to a desired functionality is probably best done during undergraduate and graduate training, and curricular modules could be developed to advance this goal.

Mainstreaming of SSA, however, will still require appropriate incentives created by government funding. Government programs for applied research and technological development, as well as health effects research, should always include assessment of alternative technological possibilities and opportunities. All major grants aimed at advancing particular avenues of technology, or advancing knowledge about the health effects of chemicals, materials, and processes, should include alternative assessment activities. Agency program managers are thus a key population of potential change agents, and government "requests for proposals" a key vehicle for incentivizing the necessary change. Importantly—and consistent with SSA principles—both the educational path and the funding incentive path require only marginal shifts in resource allocations and practitioner behavior, rather than whole-sale changes in priorities and practice.

A more ambitious policy goal would be the creation of a national Toxics Use Reduction program aimed at institutionalizing SSA across a broad cross-section of technological and economic activities. The Massachusetts Toxic Use Reduction Act (TURA) provides one successful model and shows how states can test innovations in governance. Key features of TURA include a surveillance process that identifies and lists chemicals of concern but does not directly regulate them, a focus on technological substitution rather than regulatory proscription, and support for capacity building to identify less hazardous substitutes for toxic materials. As newer areas of innovation such as nanotechnology and synthetic biology begin to bring new chemicals, materials, and processes into the world, a national-scale SSA-TURA approach to assessing and introducing these technologies could help ensure that innovation paths are more sustainable, and that the bruising political battles of past decades are defused and replaced by shared commitments to the continual discovery and exploration of a more sustainable future.

REFERENCES

1. W. Clark and N. Dickson, Sustainability Science: The Emerging Research Program, *Proceedings of the National Academy of Science USA, 100*:14, pp. 8059-8061, 2003.
2. H. Komiyama and K. Takeuchi, Sustainability Science: Building a New Discipline, *Sustainability Science, l,* pp. 1-6, 2006.

3. J. I. Bailar and A. J. Bailer, Risk Assessment—The Mother of All Uncertainties: Disciplinary Perspectives on Uncertainty in Risk Assessment, in *Uncertainty in the Risk Assessment of Environmental and Occupational Hazards,* New York Academy of Sciences, New York, pp. 273-285, 1999.
4. R. Benedick, *Ozone Diplomacy: New Directions in Safeguarding the Planet.* Harvard University Press, Cambridge, Massachusetts, 1998.
5. J. Maxwell and F. Briscoe, There's Money in the Air: The CFC Ban and Dupont's Regulatory Strategy, *Business Strategy and the Environment, 6,* pp. 276-286, 1997.
6. E. A. Parsons, *Protecting the Ozone Layer: Science and Strategy,* Oxford University Press, Oxford, United Kingdom, 2003.
7. M. M. Quinn, T. P. Fuller, et al., Pollution Prevention—Occupational Safety and Health in Hospitals: Alternatives and Interventions, *Journal of Occupational Environment and Hygiene, 3*:4, pp. 182-193; quiz D45, 2006.
8. M. Rossi, J. Tickner, et al., Alternatives Assessment Framework of the Lowell Center for Sustainable Production, Lowell Center for Sustainable Production, Lowell, Massachusetts, 2006.
9. S. Funtowicz and J. Ravetz, *Uncertainty and Quality in Science for Policy,* Kluwer Academic, Dordrecht, Netherlands, 1990.
10. D. Kriebel, J. Tickner, et al., The Precautionary Principle in Environmental Science, *Environmental Health Perspective, 109*:9, pp. 871-876, 2001.
11. D. Sarewitz, How Science Makes Environmental Controversies Worse, *Environmental Science and Policy, 7,* pp. 385-403, 2004.
12. D. Kriebel, How Much Evidence is Enough? Conventions of Causal Inference, *Law & Contemporary Problems, 72*:1, pp. 121-136, 2009.
13. M. Gibbons, Science's New Social Contract with Society, *Nature, 402*(Suppl.), pp. C81-C84, 1999.
14. TURI, *TCE Reduction Resources,* February 8, 2008. Retrieved August 13, 2008, from http://www.turi.org/laboratory/trichloroethylene_tce_reduction_resources
15. D. Edgerton, *The Shock of the Old: Technology and Global History Since 1900,* Oxford University Press, Oxford, United Kingdom, 2007.

Contributors

LENORE AZAROFF is Research Professor in the Center for the Promotion of Health in the New England Workplace, University of Massachusetts, Lowell Massachusetts.

DAVIS BALTZ is Director of the Precautionary Principle Project at Commonweal, in Bolinas, California.

DANIEL CHIA is Senior Adviser, California State Legislature, Sacramento, California.

BARRY COMMONER is former Director of the Center for the Biology of Natural Systems at Queens College in New York City.

CATHY CRUMBLEY is Program Director, Lowell Center for Sustainable Production, University of Massachusetts, Lowell Massachusetts.

GREGORY DELAURIER is a consultant with the Department of Work Environment, School of Health and Environment, University of Massachusetts, Lowell Massachusetts.

JOHN DEMENT is Program Director in Epidemiology in the Division of Occupational and Environmental Medicine at Duke University School of Medicine in Durham, North Carolina.

BRYAN EHLERS is with the California Policy Research Center, University of California, Berkeley California.

SCOTT FULMER is Program Director of the Construction Occupational Health Program in the Department of Work Environment in the School of Health and Environment at the University of Massachusetts, Lowell Massachusetts.

ROBERT HALE is Professor of Marine Science in the Department of Environmental and Aquatic Animal Health at the Virginia Institute of Marine Science, The College of William and Mary in Gloucester Point, Virginia.

SVEN OVE HANSSON was an advisor the Swedish government and is Professor and Chair of the Department of Philosophy and History of Technology at the Royal Institute of Technology, in Stockholm, Sweden.

JOSEPH HERMAN was with the Physicians for Social Responsibility.

ROBERT HERRICK is Senior Lecturer on Industrial Hygiene at the Harvard School of Public Health in Boston, Massachusetts.

POLLY HOPPIN is Research Professor and Program Director, Environmental Health, Lowell Center for Sustainable Production, University of Massachusetts-Lowell, Massachusetts.

GENEVIEVE HOWE was at the Boston University School of Public Health and is a consultant at the Ecology Center in Ann Arbor, Michigan.

ANNETTE HUDDLE was with the Physicians for Social Responsibility.

MOLLY JACOBS is Project Manager, Lowell Center for Sustainable Production, University of Massachusetts, Lowell, Massachusetts.

DAVID KRIEBEL is Professor and Chair of the Department of Work Environment in the School of Health and Environment and Co-Director of the Lowell Center for Sustainable Production at the University of Massachusetts, Lowell, Massachusetts.

MARK LaGUARDIA is a Research Scientist at the Virginia Institute of Marine Sciences, The College of William and Mary in Gloucester Point, Virginia.

CHARLES LEVENSTEIN is Professor Emeritus, Department of Work Environment, School of Health and Environment, University of Massachusetts, Lowell, Massachusetts.

JOCK McCULLOCH teaches in the School of Global Studies at Royal Melbourne Institute of Technology in Melbourne, Australia.

SUSAN MOIR is Director of the Labor Resource Center at the University of Massachusetts, Boston Massachusetts.

MARGARET QUINN is Professor in the Department of Work Environment, School of Health and Environment, University of Massachusetts, Lowell, Massachusetts.

KATHLEEN REST is Executive Director of the Union of Concerned Scientists in Cambridge, Massachusetts.

DAVID RICHARDSON is Associate Professor of Epidemiology in the Department of Epidemiology at the University of North Carolina School of Public Health in Chapel Hill, North Carolina.

DANIEL SAREWITZ is Professor of Science and Society and Co-Director of the Consortium for Science Policy, Arizona State University, currently in Washington, DC.

ELLEN SILBERGELD is Professor in the Department of Environmental Health Sciences, Health Policy and Management and Epidemiology at the Johns Hopkins Bloomberg School of Public Health in Baltimore, Maryland.

MICHAEL SILVERSTEIN was formerly with the United Auto Workers Health and Safety Department and is now Assistant Director of the Division of Occupational Safety and Health for the State of Washington Department of Labor and Industries.

GINA SOLOMON is Senior Scientist at the Natural Resources Defense Council and Associate Clinical Professor of Medicine at the University of California , San Francisco School of Medicine in San Francisco, California.

ALICE STEWART is now deceased, but was with the Department of Social and Preventive Medicine at Oxford University and was an honorary member of the Department of Social Medicine at the University of Birmingham, England.

JOEL TICKNER is Director and Associate Professor of Community Health and Sustainability, University of Massachusetts, Lowell, Massachusetts.

GEOFF TWEEDALE is a Reader in Business History at the Manchester Metropolitan University in Manchester, England.

MICHAEL WILSON is Research Scientist, University of California, Berkeley School of Public Health, Center for Occupational and Environmental Health, Program in Green Chemistry and Chemicals Policy, Berkeley, California

STEVE WING is Associate Professor in the Department of Epidemiology at the University of North Carolina School of Public Health in Chapel Hill, North Carolina.

Index